ELEMENTS OF CONSTRUCTION FOR ELECTRO-MAGNETS

by

Théodore Achille L. Du Moncel

Merchant Books

ELEMENTS OF CONSTRUCTION

FOR

ELECTRO-MAGNETS.

By COUNT TH. DU MONCEL,

MEMBRE DE L'INSTITUT DE FRANCE.

TRANSLATED FROM THE FRENCH

BY

C. J. WHARTON.

Merchant Books

PREFACE.

In presenting the accompanying translation of the 'Détermination des Eléments de Construction des Electro-Aimants' by Count Th. du Moncel I need offer no apology, as the researches of this eminent French scientist are well known both in England and America, but to facilitate the studies of those not understanding French I have undertaken the translation which I now put forward as a faithful copy of the original, but in an English dress. The electrical units are almost throughout based on the French system of measurements of a Daniell as the unit of E.M.F. and the metre of telegraph wire as the unit of resistance, but to calculate these values according to the mode now usually obtaining here, viz. the volt and the ohm, the results must be divided by 1·079 and 100 respectively.

With respect to the tables of values, I should mention I have omitted some given by the author containing information as to the weight, price, &c., of copper wires supplied by a Paris maker.

In conclusion, I am requested by Count Th. du Moncel to explain that the experiments undertaken by him in order to determine the ratios of magnetic moment for the standard electro-magnet have not been numerous and varied enough for them to be taken as presenting great exactitude. Further experiments in this direction would therefore be of advantage, and it must also be borne in mind that owing to the very varying nature of the iron employed in electro-magnets we can never be certain that the real results correspond with the theoretical. What, however, may be taken as certain is, that electro-magnets constructed and used in accordance with his formulas are employed in the most advantageous manner possible.

C. J. WHARTON.

8 AND 9, HOLBORN VIADUCT,
January 1883.

Merchant Books

CONTENTS.

CHAPTER V.

CHAPTER VI.

CHAPTER VII.

CHAPTER VIII.

CHAPTER IX.

Merchant Books

M.B

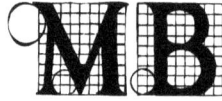

ELEMENTS OF CONSTRUCTION

FOR

ELECTRO-MAGNETS.

———•———

INTRODUCTION.

It is generally complained, and not without reason, that the manner in which the question of electro-magnets is ordinarily treated by scientists is so undecided, and so little practical in the deductions which are given, that it is impossible for a constructor or an inventor to profit by them. It is certain that mathematicians look upon the questions involved from too great a height to occupy themselves with the details of application; and it must also be admitted that magnetic theories are so vague that it is difficult to translate into mathematical symbols many of the laws which have been pointed out, and of which some are still contested by punctilious spirits.

We, who have made many experiments with electro-magnets, are less sceptical, for although we may not have been able to verify with extreme exactness the laws propounded by Lenz, Jacobi, Dub, and Muller, we have found results approaching near enough to

B

the truth to be accepted as data for the construction of electro-magnets. The laws of Ohm for electric currents are in the same predicament, for it is difficult to bring into formulas which represent them a crowd of secondary influences which derange more or less the effects enunciated. But these laws are useful guides, and serve as premises for correct deductions, and this is the essential point.

For scientific data to be of any real use in their application they must be cleared of all the hypotheses of high science, and of terms which many electricians cannot understand; and further, we must start from experiences obtained under ordinary conditions of application. It is certain that if, to appreciate a magnetic force, we are obliged to base our calculations, through complicated formulas, on the oscillations of a magnetised needle or on the currents induced by this force, constructors would say that this means nothing to their mind, and that, in fact, they know nothing of magnetic force but that which corresponds with some weight lifted or supported; in short, their desire being to have the greatest possible power under the given conditions, the rest is of no consequence to them. It is, then, under these conditions that we must state the question to obtain deductions applicable in practice. Now, I have always admitted that the known laws of electro-magnets were sufficient to satisfy constructors in this respect, and it is this which has led me to publish my various pamphlets on the best methods of constructing electro-magnets. Wishing to be certain of my deductions, I have been obliged to make numerous experiments to verify my formulas, and it is only after the most minute experimental researches that I have laid down those given in the present work.

CHAPTER I.

FORMULAS FOR ELECTRO-MAGNETS.

To establish my formulas I have started from the different elements entering into the construction of an electro-magnet—viz. the dimensions of the magnetic core, the size of the wire in the coils, its length, the number of turns, and its depth, as follows :—

a. The depth of the magnetising coils ;

b. The total length of the bobbins or the two arms of the electro-magnets ;

c. The bore of the bobbins, or what we may suppose to be identical, the diameter of the magnetic core ;

g. The diameter of the wire of the coils, including the insulating covering ;

A. The attractive force of the magnetic system ;

E. The electro-motive force of the battery employed ;

F. The magnetic moment of the electro-magnet ;

H. The total length of the wire in the coils ;

I. The intensity of the current in the total circuit ;

t. The number of turns of the wire ;

R. The resistance of the exterior circuit, including that of the battery.

It is easy to understand that we can at once represent the number of turns in each layer by $\frac{b}{g}$, and as there are as many layers as g is contained in the depth a, we shall have to represent the total number of turns t,—

$$(1) \quad t = \frac{b}{g} \times \frac{a}{g} = \frac{a\,b}{g^2}.$$

If we want to arrive at the length of each of these turns in the first or in the last layer, we shall find that for the first we have $2\,\pi\,\frac{c+g}{2}$, and for the last $2\,\pi\,\frac{c+2\,a-g}{2}$, and consequently the total length of these two layers will be

$$\frac{b}{g}\,2\,\pi\,\frac{c+g}{2} \text{ and } \frac{b}{g}\,2\,\pi\,\frac{c+2\,a-g}{2}.$$

The intermediate layers constitute, with these two, the terms of an arithmetical progression, of which the foregoing expressions are the extremes, and of which the number of terms is represented by $\frac{a}{g}$; the total length of the wire, or the sum of the lengths of these different layers, will be represented by the formula—

$$(2) \quad H = \frac{b}{g}\,\frac{2\,\pi\,(c+g+c+2\,a-g)}{4}\,\frac{a}{g} = \frac{\pi\,b\,a\,(a+c)}{g^2}.$$

We thus obtain, then, the values of t and H in terms of the different elements entering into the construction of an electro-magnet. These formulas are also useful in themselves, for they enable us to calculate the length and number of turns in the bobbins of any electro-magnet when we have measured the thickness of the wire and the depth of the coils, which are easy to obtain, since it is only necessary to measure the exact length of the bobbin from one edge to the other, to count the number of wires in this space, divide this length by the number, and take the difference in diameter of the bobbin covered with wire and the iron core. We can also deduce from these formulas the values of a and g, and have different expressions for the values of t and H, according to any particular case.

To obtain the expression of the electro-magnetic force, I start from the laws of Jacobi, Dub, and Muller, who give for the moment F of an electro-magnet, the product of the intensity I of the current traversing it by the number of turns, and as the value of the attractive force A, the square of this product, which leads to the formulas—

$$F = \frac{E\,t}{R + H} \text{ and } A = \frac{E^2\,t^2}{(R + H)^2} \cdot$$

And applying to t and H the values already arrived at we obtain the equations—

$$F = \frac{E\,a\,b}{R\,g^2 + \pi\,b\,a\,(a + c)} \text{ and}$$

$$A = \frac{E^2\,a^2\,b^2}{[\,R\,g^2 + \pi\,b\,a\,(a + c)\,]^2}$$

which allow us to deduce different conditions of maximum according as we vary the quantities $a, b, c,$ and $g,$ and which relate: Firstly, to the resistance which we give to the magnetising coils; Secondly, to the proportion which exists between these coils and the diameter of the magnetic core; Thirdly, to the dimensions themselves of the electro-magnet.

It is, however, important to remember that, to obtain the reduced length of the different parts which compose the circuit, and which are represented by very different conductors, it is necessary to transform the resistance R into terms of that of the coils; and for this we shall begin by considering that, as g represents the diameter of the wire, including the insulating covering, we must divide g by a coefficient f to obtain the diameter of the wire itself, which is all that ought to be taken into consideration. This coefficient may be represented practically by $1 \cdot 6$ for the very fine wires, or by $1 \cdot 4$ for thicker ones. The diameter of the conducting wire will then be $\dfrac{g}{f}$, and if we take q to represent the relative conductibility of R and H, including the constant for unit section, which is $\cdot 016$ mm., we shall have for the reduced value of R the quantity—

$$\frac{q \, \mathrm{R} \, g^2}{f^2}.$$

Since, by varying the size of the wire $\dfrac{g}{f}$ for a constant depth a of the coils, we vary not only its resistance but also its length—two properties which vary in the same proportion—the quantity H taken as representing the resistance of the coil instead of being in inverse proportion to g^2, will be in inverse proportion to g^4; but the quantity $\dfrac{q\,\mathrm{R}\,g^2}{f^2}$, which is not in the same case as H, because it is not obliged to fill a fixed space, will remain in inverse proportion to g^2; so that the denominators of the expressions F and A becoming

$$\frac{q\,\mathrm{R}\,g^4 + \pi\,b\,a\,(a+c)\,f^2}{f^2\,g^4}$$

and

$$\left[\frac{q\,\mathrm{R}\,g^4 + \pi\,b\,a\,(a+c)\,f^2}{f^2\,g^4}\right]^2,$$

the expressions themselves will be

$$(3) \quad \begin{cases} \mathrm{F} = \dfrac{f^2\,g^2\,\mathrm{E}\,a\,b}{q\,\mathrm{R}\,g^4 + f^2\,\pi\,b\,a\,(a+c)}, \\[2ex] \text{and} \\[2ex] \mathrm{A} = \dfrac{f^4\,g^4\,\mathrm{E}^2\,a^2\,b^2}{\left[q\,\mathrm{R}\,g^4 + f^2\,\pi\,b\,a\,(a+c)\right]^2}. \end{cases}$$

Merchant Books

CHAPTER II.

CONDITIONS OF MAXIMUM FOR ELECTRO-MAGNETS ON A SIMPLE CIRCUIT.

1st. *Conditions of maximum in reference to the resistance of the coils.*—In the preceding formulas the values of A and F may be disputed from several points. We may ask what would be the conditions of maximum, admitting that, having an electro-magnet already constructed with fixed dimensions, we wish to employ it most advantageously with an exterior circuit whose resistance is R; or we may ask—not being limited in respect to the depth of the coils—to what extent it would be advantageous to wind on the bobbins a wire of a given size, to produce, with the given resistance R, the best results? In the first instance, the variation is in the diameter g of the wire, in the second in that a of the bobbin.

If we consider that in the preceding formula the diameter of the conducting wire of the coil is not g but $\frac{g}{f}$, in which f is a constant, we find that the calculation is not so simple as would be imagined at first sight; and therefore those who first went into this question considered the value of f could be neglected, and reasoned on the hypothesis that g represented the thickness of the wire itself. Even in the preceding formulas, taking into account the value of f, this value cannot be considered constant if we take g as variable, and it will at once be seen that the conclusions will be different. But in considering the question under its most simple conditions, we are able to show with ease, by bringing in the results of the two preceding formulas, that the conditions of maximum will be found in the equation—

$$\frac{q\,\mathrm{R}\,g^2}{f^2} = \frac{\pi b a\,(a+c)}{g^2},$$

that is to say, $\mathrm{R} = \mathrm{H}$, which means that, for electro-magnets of similar dimensions having bobbins of the same diameter, the most advantageous size of wire for the coils is that which will render their resistance equal to that of the exterior circuit R.

If we take into account the thickness of the insulating material, the calculation will show that the best coil will be that of which the resistance will be to the resistance of the exterior circuit, as the diameter of the wire itself is to that of the same wire, including its insulating covering.

If we vary the depth a of the coils, supposing the action of the turns to be about equal, which we may admit under the ordinary conditions of electro-magnets, taking into account the differences of resistance wrought by their greater or less distance from the magnetic core, the conditions of maximum corresponding with the cancelment of the results of the preceding expressions as far as concerns a, will show that R should be equal to $\dfrac{\pi b a^2}{g^2}$, that is to say, to the length H of the wire of the coils divided by $\dfrac{a+c}{a}$, or, what comes to the same thing, that H should be equal to $R\left(1+\dfrac{c}{a}\right)$. Translated into ordinary language, this deduction means that:— Between several bobbins wound with the same wire but having a different number of layers, that which will give the best results with a given circuit of resistance will be that of which the resistance is to the exterior resistance, as the depth of the coils, increased by the diameter of the magnetic core, is to the depth of the coils alone.

As usually in electric applications we start with a given diameter for the magnetic core, and as for other conditions of maximum of which we will treat presently, the depth of layers of wire should be equal to the diameter of this core ; as, on the other hand, the depth of the insulating layers varies, and is undetermined in the researches which we have to make, it is the first of these conditions which ought to be considered in the construction of electro-magnets. But for one who wishes to ascertain what is the resistance of the circuit on which he can most usefully employ a given electro-magnet, it is the second conditions of maximum with which he is concerned, and these conditions show that this exterior resistance should be less than half that of the electro-magnet, if the depth of the coils is equal to the diameter of the magnetic core, which ought to be the case.

2nd. *Conditions of maximum relating to the proportion which should exist between the depth of the magnetic coil, and the diameter of the iron cores.* — A second very important point in the construction of an electro-magnet is to know what ought to be the depth of the magnetic bobbins, to make the best use of them. It will be understood that the force of electro-magnets increasing with the diameter of the magnetic cores, and the resistance of the turns of the coil becoming greater in consequence of this increase of diameter, there must be a limit where the advantages obtained from the increasing of the diameter are counterbalanced by the increase of resistance of the coil; and the question is to determine this limit. Calculation furnishes the means for resolving this question.

In the equations (3) expressing the values F and A, the magnetic moment of the electro-magnet and its attractive force, let us vary the quantity c which represents the diameter of the magnetic core, and establish between this quantity and the depth a of the coil an algebraic relation, which is easy, for, supposing the coil to be wound on the core itself, its interior diameter is represented by c. We could then, by fulfilling the conditions of maximum with respect to the resistance of the exterior circuit, obtain an expression susceptible of maximum such that the relation of R to H shall be either of those given in the first two deductions we have drawn. In representing by λ the coefficient by which the length of the coil must be multiplied to place the total circuit in either of these conditions of maximum, and supposing the depth a of the coils, and therefore the number t of the turns, to be invariable, the attractive force A and the magnetic moment F of the electromagnet are expressed, according to the law of Muller relating to the increase of force with the diameter of the magnetic cores, by

$$(4) \qquad F = \frac{g^2\, E\, \sqrt{c}}{\lambda\, \pi\, b\, a\, (a + c)}, \quad \text{and}$$

$$A = \frac{g^4\, E^2\, c}{[\lambda\, \pi\, b\, a\, (a + c)]^2},$$

expressions which are susceptible of maximum in reference to c; but then the quantities R and H are supposed to vary at the same time, in proportion as the coil is lengthened by reason of the increase of the magnetic core. If we take the results of the preceding expressions as regards c considered as variable and reduce them to zero, we find that the maximum conditions answer to $a = c$, that is to say, to the equality of the depth of the coils and of the diameter of the iron core. This is what I have demonstrated by experiment, as we shall see at the end of this work.

Consequences of the preceding Laws.—The advantages of the laws which we have propounded above are easy of comprehension, for they assist in simple calculations for the construction of electro-magnets.

By this means, in fact, the expression giving the length of the coils becomes for the two conditions of maximum which we have considered $\dfrac{2 \pi b c^2}{g^2}$, and if we take the length b of the electro-magnet in terms of the diameter c, multiplying this by the coefficient m, which practice and the considerations we are about to discuss fix at 12 for the two arms of the electro-magnet, this expression becomes—

$$(5) \quad \frac{2 \pi c^3 m}{g^2} \quad \text{or} \quad \frac{75 \cdot 4 \times c^3}{g^2},$$

formulas in which we have only to consider the two quantities c and g, which can be determined according to the different conditions in which we may find ourselves, by means of the proportion which we will consider presently.   On the other hand, we have for the number of turns t, the equation—

$$(6) \quad t = \frac{12 c^2}{g^2}.$$

Where the electro-magnet fulfils the conditions of maximum, $R = H$, $a = c$, $b = c\, m$, as g is not determined, R must be reduced to terms of g, and consequently the equation $\dfrac{2 \pi c^3 m}{g^2}$, which represents H, and which must be equal to R, leads to the equation $\dfrac{q\, R\, g^2}{f^2} = \dfrac{2 \pi c^3 m}{g^2}$ whence we have

$$g^4 = f^2 \frac{c^3}{R} \frac{2 \pi m}{q}.$$

Since $\dfrac{2\pi m}{q}$ is a constant composed of known quantities and equal to $0 \cdot 00020106$ we have as final*—

$$(7) \qquad g = \sqrt{f\sqrt{\frac{c^2}{R}} \cdot 00020106}.$$

3rd. *Conditions of maximum as regards the length of the iron core.*—From the deductions which we have just considered, it will easily be understood how important it is to calculate the lengths of the magnetic cores in terms of their diameters; but can a definite length be fixed by calculation? This is the question which remains to us to elucidate. It is certain that if we consider purely and simply the length of the magnetic core b, no maximum conditions can be deduced from the formulas giving the values of F and A; for although, according to the laws of Muller, the forces increase as the square root of the length of the magnetic cores, these formulas are not susceptible of a maximum when we vary b. But if we render this quantity b in terms of the diameter c the attractive force becomes

then proportional to $c \times \sqrt{c\,m}$ or to $c^{\frac{3}{2}}$, and we then

---

* In this expression q represents the relative conductibility of R and H, R being expressed in metres of telegraph wire. As this wire is of iron, and that of the coils of copper, this ratio is about 6. And as the diameter of this telegraph wire is 4 millimetres, the square of its section representing unity is $\cdot 000016$ [metre; thus $q = \dfrac{6}{\cdot 000016} = 375000$. Therefore the constant $\dfrac{2\pi m}{q} = \dfrac{2 \times 3 \cdot 1416 \times 12}{375000} = \cdot 00020106$.

obtain for the attractive force A, all the other conditions being taken into consideration, the expression—

$$(8) \quad A = \frac{E^2 \, m^2 \, c^4 \, c^{\frac{1}{2}}}{[R g^2 + 2 \, \pi \, c^3 \, m]^2};$$

and here R is represented by a given length of the wire of the coil.

Now, according to a simple reasoning, we may believe that there is, in that case, a limit to the value of m, for the magnetising coil having a given resistance in relation to that of the exterior circuit, and that coil having to possess a thickness equal to the diameter of the magnetic core, this resistance may be more or less completely utilised, according to the relation existing between the diameter and the length of the core upon which the coil is wound. As the electro-magnetic force increases with the diameter of that core, it is best, up to a certain point, to take it as great as possible; but on the other hand, as the number of turns, for a given length of coil, diminishes with that greatness, it may be preferable not to further increase the diameter, but to lengthen the core, and this advantage, added to the greater number of turns thus obtained, may counterbalance advantageously, in certain conditions, the loss of force resulting from the lessening of the diameter. However, theoretical formulas do not precisely define that limit, because the maximum conditions which may be deduced from the preceding formula, taking *c* as variable, answer to the equation—

$$\frac{2\,\pi\,c^3\,m}{g^2} = 11\,\mathrm{R};$$

that is to say, that according to the hypothesis which has been admitted, we can increase the dimensions of the magnetic core until the resistance of the magnetising coil represents eleven times the resistance of the exterior circuit. Under these conditions we have $m = 11\,\dfrac{\mathrm{R}\,g^2}{2\,\pi\,c^3}$, and we see that the value of *m* then becomes equal to eleven times the proportion between R and the resistance of a coil having its magnetic core equal in length to the diameter of the electro-magnet; for in this case this coil is expressed by $\dfrac{2\,\pi\,c^3}{g^2}$; now, as the proportion of the resistance R to that of the coil ought to be always 1—for these two resistances must always be equal to meet the conditions of maximum previously laid down—we get the preceding formula $m = 11$. Numerous calculations that I have given in my 'Recherches expérimentales sur les maxima électro-magnétiques' show the exactitude of these deductions.*

* Suppose, for example, introduced in a circuit of 64 metres of telegraph wire, three bar electro-magnets with diameters of 8, 7, and 6 millimetres, with lengths of eleven times these diameters, namely, 8·8, 7·7, and 6·6 centimetres. If we compare the results obtained, using wire of ·6 millimetre, we shall find the first magnet A will have a coil of 99 metres in length and 1056 metres in resistance; the second magnet B will have a coil of 66 metres in length,

As the ends of the bobbins must have a tangible thickness, and therefore the poles protrude a little beyond the coils, and there must be a space between the bottom of the ends and the breech of the magnetic core to allow of the egress of the two ends of the wire of the coil, the coefficient 11 must be a little increased, and we have found by experiments that it should be made 12.

---

and 704 in resistance; and the third magnet C a coil of 41·36 metres, and 441 metres in resistance. The number of turns deduced from the formula $\dfrac{11\,c^2}{g^2}$ will be 1955 for A, 1497 for B, 1100 for C, and the $\sqrt{\text{cube}}$ of the diameters being ·000715, ·000585, and ·000464, we have for the forces of the three electro-magnets, ·247868, ·252891 and ·250307. It will be seen from this that the electro-magnet B has the advantage, and has a resistance of 704 metres, or eleven times that of the exterior circuit. But this is no longer the case if we suppose R = 704 metres: the relative forces are then ·10037, ·07524, and ·04871, and it is the large magnet which has the advantage because it has a resistance great enough, compared with B, to prevent the latter exercising sufficient influence to greatly modify the value of $(R + H)^2$ which divides the product $t^2\,c^{\frac{4}{3}}$.

Merchant Books

CHAPTER III.

CONDITIONS OF MAXIMUM ON COMPOUND CIRCUITS.

THE deductions that we have drawn in the preceding chapter suppose the electric current to be constant and uniform, the reactions of extra currents not to exist, the iron of the electro-magnet to be in that state of magnetic saturation necessary for the laws of Dub and Muller to be applicable, and finally, the exterior circuit to be perfectly insulated. Except under these conditions the formulas will not exactly apply, and calculations prove that the resistance of the coil must be considerably reduced, as is indisputably shown by the experiments of Hughes, and confirmed by those of Lenoir made between Paris and Bordeaux; in which the electro-magnet experimented on underwent very rapid magnetisation and demagnetisation.

With such a diversity of elements it is impossible to fix, for telegraphic electro-magnets, a formula which will give exactly the conditions of maximum for the resistance of magnetic coils; but on the basis of the preceding formulas, and admitting, as experience has proved, that the electro-magnetic attractions increase in a much more rapid proportion than that of the square of the intensity of the current, when the state of magnetic saturation is not attained, we may at once assume, as we shall prove further on, that to very rapidly alternately magnetise and demagnetise, the resistance in the coils must be considerably diminished; but this diminution is again considerably augmented in a great measure by the division of the current along the lines, so that it may become necessary to reduce to 40 kilometres the resistance of an electro-magnet for telegraphic purposes in use between Paris and Bordeaux, which is about 500 kilometres. All the deductions that we have given above are, however, at fault for this application of electro-magnets; but it is not the same with a mechanical application where the circuit is well protected and where the magnetism is free to develop itself in the iron core, and where we may choose the iron to correspond with the point of magnetic saturation, such as is defined by Muller. We shall have occasion presently to inquire into these properties of the magnetic core; but to dispose of the laws we have already propounded, we have yet to examine the conditions of maximum in connection with an electro-magnet introduced in one part of a divided circuit.

Taking the most simple case, that of a single branch u with a resistance l and a common resistance R on leaving the battery, we shall find that the attractive force A of the electro-magnet on l will be

$$A = \frac{E^2\,u^2\,t^2}{[R\,(u + l + H) + u\,(l + H)\,]^2},$$

and if we substitute for t and H their true values obtained from the equations we have previously given, we get an expression whose result, in reference to g taken as variable, becomes zero when we have

$$(9) \qquad \frac{q\,g^2}{f^2}\Big(l + \frac{R\,u}{R + u}\Big) = \frac{\pi\,b\,a\,(a + c)}{g^2},$$

an equation which consequently answers to the conditions of maximum.

In varying the depth a of the coil, the quantity g remaining constant, these conditions of maximum are represented by

$$\frac{q\,g^2}{f^2}\Big(l + \frac{R\,u}{R + u}\Big) = \frac{\pi\,b\,a^2}{g^2}.$$

Now, in the first of these equations, the second side represents the resistance of the wire of the coil and the first the total resistance of the exterior circuit expressed in similar units to those of which we have made use to estimate the length of the wire of the coil. But this total resistance is taken in an inverse sense, for that which we are considering is represented, in fact, by

$$R + \frac{l\,u}{l + u}.$$

In this case the total resistance must be supposed as if the part common to the two branch currents were represented by l, and as if the part R really common were but a branch.

In the second equation the first side represents as before the total resistance of the circuit taken inversely; but this total resistance, like the resistance R of a single line, must be considered as before to be less than that of the electro-magnetic coil in the proportion of 1 to $1 + \dfrac{c}{a}$, to agree with the conditions of maximum as regards the variable quantity a.

In short, we may conclude that the laws of electro-magnetic maximum on divided circuits are the same as those relating to simple circuits, always supposing that the resistance R, on which they are based, is represented by the total resistance of the exterior circuit with its branches, and admitting that this total resistance is considered as if the battery were substituted for the electro-magnet in the circuit. Now, as the total resistance of a divided circuit is less than its proper resistance, the coil ought to have a less resistance than the latter. All these formulas have been verified by experiment, as we shall see further on.

When the branch circuits have a low resistance and it is possible to combine the elements of the battery as we wish, we arrive by calculation at a deduction analogous to the preceding, but in an inverse sense. It is then that voltaic combination which makes the resistance of the battery equal to the total resistance of the circuits, which furnishes the most advantageous result. But if, after having established this voltaic combination we connect the branch circuits direct to the poles of the battery, and try to obtain on each circuit the greatest effect possible, as frequently happens in electric applications, the conditions of resistance of electro-magnets thereon will be quite different from those we have previously considered. This time the electro-magnets, instead of having a resistance less than that of the battery, will have a resistance greater in proportion to the number of branches; and this is easily understood if we consider that, as we said above, the resistance of the battery equals the total resistance of these circuits, and this total resistance diminishing as they become more numerous, their individual resistance must be increased in compensation. If these circuits, with the electro-magnets that they contain, were all equal, this increase would be proportional to their number, for the formula giving the intensity of the current on each would be for an ordinary battery $\dfrac{n\,\mathrm{E}}{x\,n\,\mathrm{R} + \mathrm{H}}$, x representing the number of the circuits; but if they are of unequal resistance, and if we represent by b the number of elements joined up for quantity in each group, and by a the number of groups joined up for tension, the preceding formula will be for two branches u and H

$$\frac{a\,\mathrm{E}}{\dfrac{a}{b}\,\mathrm{R}\left(1 + \dfrac{\mathrm{H}}{u}\right) + \mathrm{H}},$$

and the conditions of maximum of this formula in reference to *a* will answer to

$$\frac{a}{b} R \left(1 + \frac{H}{u} \right) = H, \text{ or } H = \frac{a R u}{u b - a R},$$

an equation which becomes $H = 2 \frac{a}{b} R$ when $u = H$.

We shall see later an application of these principles.

CHAPTER IV.

APPLICATION OF THE LAWS OF MAXIMUM TO THE CONSTRUCTION OF ELECTRO-MAGNETS.

THE different laws and formulas that we have just enunciated allow of an easy solution of the problems relating to electro-magnetic attractions; but for this we must call in Muller's law relating to magnetic saturation, which law establishes that, to develop in two electro-magnets the same percentage of their magnetic maximum, the intensities of their exciting currents, multiplied by the number of turns of wire, must bear the proportion one to the other of the $\sqrt{\text{cube}}$ of the diameter of each magnet. This law may be thus expressed:—

$$\frac{I\,t}{I'\,t'} = \frac{\sqrt{c^3}}{\sqrt{c'^3}}.$$

With this law we can easily understand that it is possible to calculate the conditions required to put an electro-magnet of a given or calculated diameter into a suitable state of saturation, not only to furnish all the force of which it is capable, but also that the laws of Jacobi, Dub, and Muller may be applicable to it. For this it is sufficient that two of the terms of the preceding proportion should be supplied by experiment, and the given quantities can easily be obtained by means of a standard electro-magnet, of which we may augment the magnetic power by increasing the electric intensity till we find that the force produced is as the square of the intensity. From experiments I made with an electro-magnet, the diameter of the iron core of which was one centimetre, and the resistance of the coil equal to 200 kilometres in a circuit of 118,600 metres, I found that the proportion in question could be obtained when the exciting battery was composed of twenty Daniell cells. As the dimensions of this electro-magnet were known, it was easy, by means of the preceding formulas, to establish constants to be made use of in our calculations. Thus we shall see how the preceding equation joined to the following—

$$ t = \frac{m\,c^2}{g^2}, \; \mathrm{H} = \frac{2\,\pi\,c^3\,m}{g^2}, \; \mathrm{I} = \frac{\mathrm{E}}{2\,\mathrm{R}}, \; \mathrm{R} = \mathrm{H}, $$

has solved the problem of which we have spoken by bringing back the value of the diameters c to a simple combination of the quantities E and R.

If, in the first equation given, we suppose the values with an accent known as bearing a proportion to the pattern electro-magnet, and if we substitute for the quantities I and t their value drawn from the conditions of maximum we have explained, we have

$$\frac{\sqrt{c'^3}}{I't'} \cdot \frac{E\,m\,c^2}{2\,R\,g^2} = \sqrt{c^3},$$

and as the diameter g is undetermined and must satisfy on one side $R = H$ and on the other $a = c$, it should be calculated in terms of these two quantities by means of the equation $R = \dfrac{2\pi c^3 m}{g^2}$; only as g is undetermined and consequently variable, the quantity R must be reduced to terms of g, which gives g^2 a value represented by $\dfrac{f\sqrt{2\pi c^3 m}}{\sqrt{q\,R}}$. In substituting this last value for g^2 in the preceding equation, we have

$$c = \frac{E}{f\sqrt{R}} \left(\frac{\sqrt{c^3 q\,m}}{2\,I't'\sqrt{2\pi}} \right),$$

a formula in which the quantity in parentheses is a constant which varies according to the system of measures employed, but which refers either to known quantities—as m, which should equal 12, q, which should equal 375,000, π, which is equal to $3\cdot1416$ —or to quantities considered as given, as they can be obtained from the pattern electro-magnet of which we know the conditions.

From another point of view, as in working out for the value $\dfrac{\sqrt{c'^3}}{I't'}$ the same calculations which were made for $\dfrac{\sqrt{c^3}}{\mathrm{I}t}$, we get the equation—

$$c = c'\,\frac{\mathrm{E}}{\mathrm{E}'}\;\frac{\sqrt{\mathrm{R}'}}{\sqrt{\mathrm{R}}}\;\frac{f'}{f}\,;$$

and as the proportion $\dfrac{f'}{f}$ is clearly equal to 1, we can rid the constant of the factor f' and consequently the simple formula giving the value of c becomes

$$(10) \qquad c = \frac{\mathrm{E}}{\sqrt{\mathrm{R}}}\,\mathrm{K},$$

K being a constant whose value varies with the units adopted.

If E represents the electro-motive force with a Daniell cell as unit, and if R is estimated in metres of telegraph wire, $\mathrm{K} = \cdot 172175$, and the figure obtained represents fractions of a metre.

In comparing the values of K and R with the system of electric measures established by the British Association, that is to say, the volt and ohm, $\mathrm{K} = \cdot 015957$.

The diameter c being known, all the other elements for the construction of an electro-magnet may be easily determined in conditions of maximum by means of the formulas we have given (5, 6, and 7), which represent the values of g, b, t, and H.

To know the value of the attractive force A, it is sufficient to work out the formula $\mathrm{I}^2\,t^2\,c^{\frac{1}{2}}$, admitting that $I = \dfrac{E}{2R}$, and that the values of t and c are determined by the formulas 6 and 10, and to find this force in weight we must compare these values with those of the pattern electro-magnet, which gives as constant ·0000855,* and leads to the formula—

$$(11) \quad A = \frac{I^2\, t^2\, c^{\frac{2}{3}}}{·0000855} \cdots \text{grammes.}$$

From the preceding formula important consequences result, which are as follows:—

1. For equal circuit resistances the diameter of an electro-magnet under maximum conditions must be proportional to the electro-motive force employed.

2. For equal electro-motive forces these diameters must be in inverse ratio to the square root of the resistance of the circuit, including that of the battery.

3. For equal diameters the electro-motive forces must be proportional to the square roots of the resistance of the circuit.

4. For a given magnetic force, and with electro-magnets under maximum conditions, the electro-motive forces of the exciting batteries should be proportional to the square roots of the resistance of the circuit (see 12).

* This coefficient represents the ratio of force calculated from the standard electro-magnet, represented by $I^2\ t^2\ c^{\frac{2}{3}}$ with the weight in grammes corresponding to this force, which is 26·85 grammes at a distance of 1 millimetre. We will refer to this more fully hereafter.

The preceding formulas also allow of the easy solution of many problems which frequently arise in electric applications, particularly to calculate the force of the battery and the dimensions of an electro-magnet required to furnish a given attractive force on a given circuit of resistance. It is true that the results obtained often do not correspond exactly with the calculation, by reason of the different nature of the iron, which may be more or less adapted to develop magnetic action, and of the more or less complete magnetic saturation of the iron, but we may always get approximate figures.

To resolve the problem in question, let us first see how the fourth postulate that we have just formulated may be demonstrated, and how, in this way, we can obtain a value representing the attractive magnetic force in terms of the electro-motive force of the battery and of the resistance of the circuit.

If we consider that the expression $I^2 \, t^2 \, c^{\frac{3}{2}}$, which represents this force, can be converted by successive substitutions* in the values of I, t, and c into the

---

* As follows :—

$$\text{Value of } I^2 \quad \ldots \quad \ldots \quad \ldots \quad I^2 = \frac{E^2}{4 \, R^2} .$$

$$\text{Value of } t^2 \quad \ldots \quad \ldots \quad \ldots \quad t^2 = \frac{E \sqrt{R}}{f^2} \, 123903.$$

$$\text{Value of } I^2 \, t^2 \ldots \quad \ldots \quad I^2 \, t^2 = \frac{E^3}{f^2 \, R^{\frac{3}{2}}} \, 30976.$$

$$\text{Value of } c^{\frac{3}{2}} \quad \ldots \quad \ldots \quad \ldots \quad c^{\frac{3}{2}} = \frac{E^{\frac{3}{2}}}{R^{\frac{3}{4}}} \cdot 07195.$$

$$\text{Value of } A \quad \ldots \quad \ldots \quad I^2 \, t^2 \, c^{\frac{3}{2}} = \frac{E^{\frac{9}{2}}}{f^2 \, R^{\frac{9}{4}}} \, 2228.$$

formula— $$\frac{E^{\frac{3}{2}}}{R^{\frac{3}{2}}}\frac{Q}{f^{2}},$$

Q is a constant equal to 2228.* For a like attractive force we get the proportions—

$$(12) \quad \frac{E^{\frac{3}{2}}}{E'^{\frac{3}{2}}} = \frac{R^{\frac{3}{2}}}{R'^{\frac{3}{2}}} \quad \text{or} \quad \frac{E}{E'} = \frac{\sqrt{R}}{\sqrt{R'}}.$$

As in the values of E and R, n, n' appear as the
number of cells employed, we may easily calculate
these numbers, knowing the values of the constants
e and ρ of the element employed, for, starting from
the preceding equation, we have—

$$\frac{n\,e}{n'\,e'} = \frac{\sqrt{n\,\rho + r}}{\sqrt{n'\,\rho' + r'}},$$

and if the accentuated quantities are in proportion
to those of the standard electro-magnet which are
known, it is easy to deduce from the preceding
equation the value of n; it is only an unknown
quantity of the second degree.

If, in the problem now under consideration, we
admit that the force of electro-magnetic attraction
is expressed in weight, it will be understood that the

formula $\frac{E^{\frac{3}{2}}}{R^{\frac{3}{2}}}\frac{Q}{f^{2}}$ could only represent this force, which

---

* The unit of electro-motive force is here taken as that of the
Daniell cell, and the resistances are calculated in metres of telegraph
wire.

we will call P, by applying to it a coefficient K, representing $\dfrac{F'}{P'}$ deduced from the data of the standard electro-magnet, whose value for an attraction at the distance of 1 millimetre is $\dfrac{\cdot002297}{26\cdot85}$ or $\cdot00008555$. It will be observed that the quantity F′ here represents for the standard electro-magnet, the formula $I^2\, t^2\, c^{\frac{3}{2}}$ or its equivalent, and the result will be that by stating the following—

$$\frac{n\, e^{\frac{3}{2}}}{(n\, \rho + r)^{\frac{3}{2}}}\; \frac{Q}{f^2} = P\,K,$$

we shall readily deduce the value of $n\, e$, which will be

$$n\, e = \sqrt{n\, \rho + r}\; \sqrt[3]{\sqrt[3]{\frac{f^4\, P^2\, K^2}{Q^2}}}\,,$$

and if we join in one the two constants Q and K, and then apply it, we have—

$$\frac{n^2\, e^2}{n\, \rho + r} = \left(\cdot0225\, \sqrt[3]{\sqrt[3]{f^4\, p^2}}\right)^2.$$

In representing by A the quantity in parenthesis, which may easily be calculated, we have—

$$n = \frac{A^2\, \rho}{2\, e^2} + \sqrt{\left(\frac{A^2\, \rho}{2\, e^2}\right)^2 + \frac{A^2\, r}{e^2}}\,.$$

The values of $n\, e$ and $n\, \rho$ being thus determined, it is easy to calculate the dimensions of the electro-magnet by means of the formulas which have already been given.

In the preceding calculations we suppose the resistance of the exterior circuit to be great enough to warrant the battery being joined up for tension; but when this resistance is small enough to lead us to suppose it would be advantageous to group the elements in series, the calculation will be somewhat modified. There will then be two alternatives; either the resistance R will be nil, or it will only be less than $n\ r$. In the first case, it will be impossible to obtain the individual values of the quantities a and b, representing respectively the number of series, and the number of elements in each series; for, the interior resistance of the battery (which is represented by $\frac{a}{b}\rho$) and the resistance H of the coils being two undetermined quantities which should be equal, we can give them what value we like, within the limits of course that the number of elements in tension includes the whole of the elements in the battery, and that the number of elements in quantity is similarly situated; * but we can obtain the value

---

* This deduction may readily be proved by the following calculations. (See also Chapter IX.) From what we have seen, the attractive force of an electro-magnet may be represented by I^2 H, and for batteries arranged in series, this expression becomes

$$A = \frac{a^2\ E^2\ H}{\left(\frac{a}{b}\rho + H\right)^2} = \frac{a^2\ E^2\ H}{4\frac{a^2}{b^2}\rho^2}.$$

Suppose we have 24 cells, and the resistance of each equals

of $a\,b$, or the total number n of the elements by means of the preceding formulas, which give under these conditions—

$$a\,b = \frac{\rho}{e^2} \cdot 000506 \sqrt[3]{\sqrt{P^4 f^8}}$$

or

$$a\,b = \frac{\rho}{e^2} \left(\cdot 0225 \sqrt[3]{\sqrt{p^2 f^4}} \right)^2 .$$

Under these conditions the value of c is determined and shown by the formula—

$$(13) \qquad c = \frac{e \sqrt{a\,b}}{\sqrt{\rho}} \cdot 173.$$

In the second case, since to satisfy the conditions of maximum as regards the electro-magnet, its resistance should equal $\frac{a}{b}\rho + r$, and that, to obtain the minimum value of the resistance of the simple circuit $\left(\frac{a}{b}\rho + r \right)$ corresponding to the maximum of

---

12 metres, and the electro-motive force equals 1; when all the cells are joined for tension, we shall have

$$A = \frac{1 \times 288}{4 \times 12^2} = \cdot 5 ;$$

when for quantity

$$A = \frac{24^2 \times 1 \times \cdot 5}{4 \times 12^2} = \cdot 5 ;$$

when they are in series of cells in quantity with $a = b$, we shall have

$$A = \frac{24 \times 1 \times 12}{4 \times 12^2} = \cdot 5 ;$$

all identical values.

the value I, on which resistance is based that of the electro-magnet, $\frac{a}{b}\rho$ must equal r, we shall then be able to determine the quantities a and b as soon as we have ascertained the value of $a\,b$ or n, and this quantity is determined by the preceding equation giving the value of $a\,b$, because $\frac{a}{b}\rho + r$ then equals $2\frac{a}{b}\rho$. From this we have—

$$a = \sqrt{\frac{n\,r}{\rho}} \text{ and } b = \sqrt{\frac{n\,\rho}{r}} \text{ or } \frac{n}{a}.$$

If, instead of a simple circuit, we consider a compound circuit with a number x of branches issuing directly from the poles of the battery, as frequently occurs in electric applications, and if on these circuits be introduced electro-magnets of similar resistance and dimensions, the value of the intensity of the current on each branch will be $\dfrac{a\cdot\mathrm{E}}{x\dfrac{a}{b}\rho + \mathrm{H}}$

or $\dfrac{b\,\mathrm{E}}{2\,x\,\rho}$, and the value of $a\,b$ or n will be the same as in the most simple case, only multiplied by x.

Let us now see how, the dimensions and construction of an electro-magnet being given, we can determine the number of elements and the arrangement of the battery that will best produce in this magnet the given force P on each of the x branches issuing from the battery.

According to the results then obtained, the quantity H in the equation $I = \dfrac{a\,E}{x\dfrac{a}{b}\rho + H}$ is ascertained,

and we know that for the conditions of maximum $x\,\dfrac{a}{b}\,\rho$ should equal H. On the other hand, we know that we should have $I^2\,t^2\,c^4 = P\,K$. Now as the quantities t^2 and c^4 may be easily calculated, as c is given, the value of I may be deduced from the preceding equation in which—

$$I = \sqrt{\dfrac{P\,K}{t^2\,c^4}}\,.$$

From this we have—

$$a = \dfrac{2\,I\,H}{E} \quad \text{and} \quad b = \dfrac{a\,x\,\rho}{H}\,.$$

CHAPTER V.

NUMERICAL EXAMPLES OF THE APPLICATION OF THE PRECEDING FORMULAS.

To facilitate putting into practice the formulas already given and the manner of calculating them, we will give several numerical examples.

First, we will suppose that we wish to find the best dimensions to give an electro-magnet before working it with a single ordinary Bunsen cell, on a circuit having scarcely any resistance beyond that of the coil. We will suppose that the electro-motive force of this element, taking the Daniell as unity, is 1·86 and its resistance 57 metres of telegraph wire, or ·57 ohm. These quantities show the necessity of knowing the values of the voltaic constants, and at the end of this work will be found a table of the principal batteries employed.

According to formula No. 10 we shall have for the value of the diameter of the iron core of the electro-magnet

$$c = \frac{1 \cdot 86}{\sqrt{57}} \cdot 172175 = \cdot 0424 \text{ metre.}$$

Each of the arms will thus have a diameter of 4·25 centimetres and a length of 25·5 centimetres.

The diameter g of the wire of the coil will be according to formula No. 7

$$g = \sqrt{1 \cdot 4 \sqrt{\frac{\cdot 0424^3}{57} \cdot 00020106}} = 4 \cdot 865 \text{ millim.,}$$

including its insulating covering, or $3 \cdot 36$ millimetres without this covering, supposing g to be divided by f or $1 \cdot 4$. The length of the wire will be, by formula No. 5, 243 metres, and the attractive force in weight at

1 millimetre, according to the formula, $A = \dfrac{I^2 f^2 c^4}{\cdot 0000855}$

will be $23 \cdot 112$ kilogrammes.

It is needless to say that if the quantity R is given in ohms instead of metres of telegraph wire, the constant of formula No. 7, instead of being $\cdot 00020106$, should be considered as $\cdot 0000020106$, so that the formula giving the value of g will then be

$$g = \sqrt{f \sqrt{\frac{c^3}{R} \cdot 0000020106}}.$$

The number of turns is calculated by means of formula No. 6.

Let us now suppose that this electro-magnet is to be introduced on a line of 100 kilometres, which with the resistance of the battery consisting of 20 Daniell cells, will constitute a circuit of 118620 metres of resistance in units of telegraph wire; we shall have

$$c = \frac{20}{\sqrt{118620}} \cdot 172175\ldots = 1 \text{ centimetre;}$$

or, using the English units of volt and ohm,

$$c = \frac{21 \cdot 58 \text{ volts}}{\sqrt{1186 \text{ ohms}}} \cdot 015957... = 1 \text{ centimetre}.$$

Each of the arms will then be 1 centimetre in diameter and six centimetres in length; the diameter of the wire will be $\cdot 2597$ millimetre with its covering, or $\cdot 1583$ millimetre without this covering, and its length 1116 metres. The attractive force will be $26 \cdot 85$ grammes. These figures show that electro-magnets on long circuits should be of small dimensions and covered with fine wire, and that, on the contrary, they should be very large when the circuit is short and the battery supplies electricity of quantity.

These dimensions for an electro-magnet excited by a Bunsen cell may appear surprising to many; but this would be still more the case if the iron had a less magnetic saturation. Thus a tubular electro-magnet whose diameter was 10 centimetres, the length of each arm 30 centimetres, and the thickness of the tube 1 centimetre, has been made to lift even without the addition of iron mouthpieces at the polar extremities, a weight of 160 kilogrammes, under the influence of a single Bunsen cell and of a coil of copper wire 4 millimetres in diameter (not including the insulating covering), only 182 metres in length and taking only 482 turns. Now, with an ordinary electro-magnet 2 centimetres in diameter, wound with a copper wire of 1 millimetre, I have never been able to obtain with the same Bunsen more than 12 kilogrammes of attractive force on contact.

Let us say that we wish to obtain a given force on a given circuit of resistance, and we wish to have an attractive force of 273 grammes (the armature being at a distance of 1 millimetre) on a circuit of 50 kilometres of resistance (or 500 ohms) with a battery of gravitation bichromate cells. In this battery the electro-motive force of each element is 2 (that of Daniell being 1), and the internal resistance ρ is about 1000 metres of telegraph wire (or 10 ohms). According to the formulas on p. 32, we shall have

$$A = \cdot 0225 \sqrt[3]{\sqrt[3]{1 \cdot 37^4 \times 273^2}} = \cdot 09 \text{ and } A^2 = \cdot 0081,$$

and

$$n = \frac{8 \cdot 1}{8} + \sqrt{\left(\frac{8 \cdot 1}{8}\right)^2 + \frac{\cdot 0081 \times 50000}{4}} = 11 \cdot 125,$$

whence

$$c = \frac{11 \cdot 125 \times 2}{\sqrt{61125}} \times \cdot 173 = 1 \cdot 553 \text{ centimetre.}$$

This gives, for the length of each bobbin, $9 \cdot 32$ centimetres.

For the diameter of the wire, with its covering, $\cdot 3894$ millimetre.

For the diameter of the same wire, without its covering, $\cdot 2842$ millimetre.

For the length of the said wire, 1861 metres.

For the number of turns, 19078.

For the intensity of the current, $\cdot 0001859$.

For the value of $c^{\frac{1}{2}}$, ·001935.

By squaring the values of I and t, and multiplying by $c^{\frac{1}{2}}$, we get ·0243778517, which represents the electro-magnetic force; and this value, compared with that of the standard electro-magnet which is ·002297, gives the proportion 10·6, which is very near that of the two weights 273 and 26·85 representing in grammes the attractive forces, particularly when we consider that some figures have been forced and decimals neglected to simplify the calculations.

We will now suppose that we wish to obtain separately, from six electro-magnets put in direct communication with the battery, a force of 200 grammes, employing a battery similar to the preceding. The total number of elements $a\,b$ will be given by the formula

$$ a\,b = \frac{6 \times 1000}{4} \times ·000506 \sqrt[3]{\sqrt{200^4 \times 1·4^8}} = 10·778, $$

or eleven elements; but, as we cannot group eleven elements into multiple series, we will suppose the battery to be composed of 12 elements. With this number the battery may be grouped in 3 sets of 4 elements, or two sets of six; but in any case we shall have

$$ c = \frac{\sqrt{12}}{\sqrt{1000 \times 6}} \times 2 \times ·173 = 1·53 \text{ centimetre,} $$

and consequently the length of each arm of the electro-magnet will be ·0918, or 10 centimetres.

And here are the values of g, t, I, &c., which we shall obtain in each case.

1st. Where $a = 3$, and $b = 4$:—

$g^2 = \cdot000000560075$; $\quad g = \cdot0007484$; $\quad \dfrac{g}{f} = \cdot0005346$.

H = 482 metres.

$t = 5015$; $t^2 = 25150225$.

I $= \cdot000666$; I$^2 = \cdot000000443556$.

$c^t = \cdot001892$.

I^2 $t^2 = 11\cdot155533$; I^2 t^2 $c^t = \cdot021084$.

A = 247 grammes.

2nd. Where $a = 2$, and $b = 6$:—

$g^2 = \cdot00000083986$; $g = \cdot000916$; $\dfrac{g}{f} = \cdot0006543$.

H = 321 metres.

$t = 3345$; $t^2 = 11189025$.

I $= \cdot001$; I$^2 = \cdot000001$.

$c^t = \cdot001892$.

I^2 $t^2 = 11\cdot189025$; I^2 t^2 $c^t = \cdot0211686353$.

A = 247 grammes.

We need not, however, assume that to preserve perfect magnetic saturation, and to maintain a constant force in an electro-magnet, we are obliged to vary its dimensions according to the resistance of the exterior circuit; we can, in certain applications, maintain the same type of electro-magnet very well by varying the force of the supplying battery and the size of the wire of its magnetising coil. In fact, as the electro-motive force of the battery employed, for a like diameter of electro-magnet and a like magnetic force, is in proportion to the square root of the exterior resistance of the circuit, we may conclude that if we increase the E. M. F. in proportion as the resistance of the circuit increases, we may maintain the same diameter for the electro-magnet, and the same force. Consequently, if, instead of introducing our standard electro-magnet of 1 centimetre in diameter, on a circuit of 100 kilometres, we introduce it on a circuit of 400 kilometres with a double number of elements, nearly the same effect will be produced. But the size of the wire must be changed, for this varies for a like diameter of electro-magnet in the inverse ratio of the $\sqrt{}$ of the resistance of the exterior circuit. This diameter, then, the wire of the standard electro-magnet being ·1583 millimetre, must be, in the present example,

$$\text{as } \sqrt[]{100000} : \sqrt[]{400000}, \text{ or as } 17\cdot8 : 25\cdot1,$$

which makes it ·112 millimetre. We see by this that we may without inconvenience maintain for telegraphic relays the allotted dimensions.

We have said above that by doubling the number of elements in the battery we should obtain nearly the same force on a circuit of 400 kilometres as on that of 100 kilometres. This word " nearly " must not be overlooked, for the resistance of the circuit, in the example we have cited, is not 400 kilometres, but 407210 metres, and this number is not exactly four times 118620, which represents the resistance in metres of the first circuit; it is less, and consequently, to obtain exactly the same effect in both cases, the battery of 20 elements must be increased to $38 \cdot 4$ elements, or 39 instead of 40.

We will now employ two electro-magnets of small dimensions, each having a resistance of 2200 metres, wound with No. 16 wire ($\cdot 44$ millimetre), and connected direct to a battery of eight Daniell cells arranged for tension. Experiment shows that the force obtained from each of them will be 70 grammes at a distance of 1 millimetre. Consequently the total force developed by both will be 140 grammes. Now if, instead of employing two electro-magnets we use but one, this force will be 200 grammes. Thus we lose 60 grammes of power by employing two electro-magnets instead of one, although the two are put in direct communication with the poles of the battery. We might think, at first sight, that there was some fault in the electric communications, but our calculations prove it to be otherwise.

In applying the formulas of Ohm to the two cases we find :—

1st. That in the case of a single electro-magnet, the value of the intensity of the current, the E. M. F. of the Daniell battery being taken as 5973, is $4 \cdot 95$.

2nd. That in the case in which the two electro-magnets are on two circuits, this intensity is represented by $2 \cdot 79$.

The combined forces of these electro-magnets then, arranged in these two ways, will be as the squares of the two quantities $4 \cdot 95$ and $2 \cdot 79$, and if we admit that the force of the electro-magnet in the simple circuit would be 200 grammes, the following formula will give the force of the two others :—

$$\frac{\overline{2 \cdot 79}^2 \times 200}{\overline{4 \cdot 95}^2} \text{ or } 64 \text{ grammes};$$

so the joint force of the two electro-magnets will be 128 grammes instead of 200, as produced by a single one. This is a still more marked difference than that shown by experiment. This effect is in consequence of the arrangement of the battery, which is already out of proportion to the resistance of the exterior circuit, as its resistance is nearly four times as great as the latter, being still less in proportion to the circuit composed of two branches. Indeed, the formula $\dfrac{n \, \mathrm{E}}{2 \, n \, \rho + \mathrm{H}}$, which represents in this case the intensity of the current on each branch, and which, in the case of a battery arranged in series, gives for conditions of maximum $2 \dfrac{a}{b} \rho = \mathrm{H}$ or $\dfrac{a}{b} \rho = \dfrac{\mathrm{H}}{2}$, and this shows that in this case the resistance of the battery should be half that of each circuit. Consequently, the number a of the elements in tension should be obtained by the formula $a = \sqrt{\dfrac{n \, \mathrm{H}}{2 \rho}}$, which in the present case gives

$$a = \sqrt{\frac{8 \times 2200}{2 \times 931}} = 3 \cdot 074.$$

As we cannot divide a cell into fractions, and as the couplings require numbers which are even divisors of n, the voltaic combination which will best answer to these conditions of maximum will be that in which the elements, separate and combined, will come the nearest to a, b, and n, and we shall see that, in the present case, it is that which contains three elements in tension, each composed of three elements in quantity, which will be the most advantageous. Now, under these conditions the attractive force of the single electro-magnet would be 267 grammes, and that of the two combined 316 grammes. Thus we gain by employing two electro-magnets.

We see by this how important it is that the construction of electro-magnets and the arrangement of the battery in electric applications should be preceded by calculation, and how much would be gained if the theory we are now propounding were well understood by constructors.

CHAPTER VI.

EXPERIMENTAL VERIFICATION OF THE LAWS OF ELECTRO-MAGNETS.

As I have already said, the formulas I have given have all been verified by experiment, except that generally adopted, which presents great difficulties in its proof, as we shall presently see. This experimental verification has seemed to me the more important, as originally many scientists did not accept these formulas without discussion, and besides, mathematical deductions are not always sufficient to convince. I have, therefore, undertaken several series of experiments to prove the truth of my deductions, and the following is the result :—

1st. The experimental demonstration of the principle, establishing the fact that the resistance of the electro-magnet should equal that of the exterior circuit, is rather delicate, by reason of the difficulty we meet with in obtaining in all the wires of commerce of different diameters conductibility exactly proportional to the squares of those diameters. The experiments I have made have been rather contradictory, and I may say contrary to theoretic deductions, which made me hesitate, in my early investigations, in admitting them, although they were generally accepted; however, as the other deductions have been confirmed by experience, and as the non-success of those I undertook on this matter might arise from accidental causes, I have been obliged to admit the principle, at least in cases where it is really applicable; that is to say, when, the depth and length of the bobbins being first fixed, we wish to know what diameter of wire will correspond best with a circuit of a given resistance. This is evidently the case in which we find ourselves when we have determined the elements for the construction of an electro-magnet, for we must, above all, proportion its core to the electric intensity which excites it, the depth of the coils being determined by a second principle which must not be lost sight of.

It is not the same, however, if we wish to ascertain what should be the resistance of the exterior circuit on which we could most usefully employ a given electro-magnet or galvanometer, without taking into account its dimensions. In this case the conditions of maximum that I have established show that the resistance of the electro-magnet should be to that of the exterior circuit as the depth of the magnetising coil, including that of the magnetic core, is to the depth of the coil alone.

This deduction may easily be demonstrated, and this is how I have drawn up my experiences.

I first wound with extreme care on the same bobbin, 6·1 centimetres in length between the ends and with an exterior diameter of 1·1 centimetre, two lengths of 60 metres of No. 16 wire, and another of

57·25 metres, which formed a third coil. These three coils had their ends out and distinct one from another, so that they could be worked separately or all together. The first gave a resistance of 1080 metres of telegraph wire, the second nearly the same, which constituted for both together a resistance of 2160 metres, and the third added to these made a total resistance of 3200 metres.

When I tried this magnet in my electro-magnetic balance with one pole only (that covered with the bobbin), I obtained, under the influence of a Leclanché battery of three elements, the total resistance of which was about 1200 metres, the following results :—

Resistance of Exterior Circuit in metres.	Coil A of 1080 metres.	Coil B of 2160 metres.	Coil C of 3200 metres.
	gr.	gr.	gr.
$1200 + \quad\; 0 = 1200$	$F = 112$	$F' = 122$	$F'' = 112$
$1200 + \quad 400 = 1600$	$F = \;\; 73$	$F' = \;\; 92$	$F'' = \;\; 95$
$1200 + 1000 = 2200$	$F = \;\; 47$	$F' = \;\; 66$	$F'' = \;\; 70$
$1200 + 2000 = 3200$	$F = \;\; 27$	$F' = \;\; 43$	$F'' = \;\; 50$
$1200 + 3000 = 4200$	$F = \;\; 17$	$F' = \;\; 29$	$F'' = \;\; 36$
$1200 + 4000 = 5200$	$F = \;\; 12$	$F' = \;\; 22$	$F'' = \;\; 28$

The attractive forces F, F', F'' were measured at a distance of 1 millimetre. This table shows that it is the coil B which most nearly approaches the theoretical requirements; that is to say, with a resistance of $\dfrac{2160}{1 + \dfrac{c}{a}}$, or 982 metres, which gives the maximum effects; and it is only when the resistance of the exterior circuit reaches 1600 metres

E

that is, $\frac{3200}{2}$, that the coil C, the most resisting, begins to show its preponderance.

With a battery of two elements and an exterior resistance of 800 metres represented by that of the battery, the force of the coil B still preponderates, being 60 grammes, while those of both the other coils, A and C, were but 57 grammes; but with a single cell, and consequently an exterior resistance of 400 metres, the coil A had the advantage, and the forces were: for A 21 grammes, for B 19 grammes, and for C 17 grammes.

These experiments, repeated with a galvanometer, were still more decisive, as may be seen from the experiments, the results of which I published in the 'Comptes rendus des séances de l'Académie des Sciences' of August 13th, 1877, page 379.

2nd. To demonstrate that on a divided circuit the resistance which should serve as basis for that of a given electro-magnet or galvanometer, is represented by the total resistance of the exterior circuit, taken inversely; that is to say, substituting the generator for the electro-magnet, I introduced in a very feeble circuit * a sensitive Ruhmkorff galvanometer, with two coils, the resistance of one of which was represented by 733 kilometres, and of the other by 237 kilometres. I added a rheostat to the circuit, and placed a second in a branch circuit, joining the two ends of the wire of my galvanometer. Under these conditions, taking R to represent the battery circuit,

---

* The generator was composed of an iron wire and one of copper joined together; it had an electro-motive force of $\frac{1}{15}$ of a Daniell, and a resistance of 272 kilometres of telegraph wire.

u the galvanometer branch, and l the branch in which the galvanometer was introduced, the equation giving the maximum conditions was

$$l + \frac{R\,u}{R + u} = \frac{\pi\,b\,a^2}{g^2}.$$

It is true l could not then figure in the calculations, for the points of bifurcation of the two branches corresponded with the two ends of the galvanometer wire; but, by means of the two rheostats, I could vary the resistances u and R so as to form a total resistance, starting from the galvanometer, which was equal to, less or greater than, the resistance of the less resisting galvanometer coils; and for this it was sufficient, R being given, to calculate u by means of the formula $u = \dfrac{R\,H}{R - H}$, H representing the resistance of the coil. Below are the results I obtained, taking all precautions not to vary the conditions during the experiments.

Circuits.	Total Resistance.	Coil of 237 kilometres.	Coil of 733 kilometres.
R = 512 + 272 u = 86	kilometres 78	$I = 27\frac{1}{2}°$	$I' = 24\frac{1}{4}°$
R = 512 + 272 u = 128	110	$I = 32\frac{1}{4}$	$I' = 30$
R = 512 + 272 u = 200	160	$I = 36\frac{1}{4}$	$I' = 36\frac{1}{4}$
R = 512 + 272 u = 256	193	$I = 40$	$I' = 41$
R = 512 + 272 u = 512	309	$I = 46$	$I' = 51$
R = 256 + 272 u = 200	145	$I = 36$	$I' = 33\frac{1}{4}$
R = 256 + 272 u = 512	260	$I = 47$	$I' = 52$

We see by this table that, as I had stated, the high-resisting coil preserves the advantage, even when the total resistance of the circuit, outside the galvanometer, is nearly equal to that of the less-resisting coil, and it is only when this total resistance falls below 160 kilometres, or 145 kilometres (according to the value of R), the point where the two coils have the same sensitiveness, that the superiority of the less-resisting coil begins to show itself. As with the galvanometer experimented on, the proportions of the two resistances (those of the exterior circuit and of the galvanometer) should be, according to the conditions of maximum deduced from the proportion as $1\cdot9 : 2\cdot425$; the resistances of the exterior circuit on which we could most usefully employ the two coils would be 98 kilometres for the low resistance and 386 kilometres for the high resistance one. And the point where they become equal is the mean proportion between these two numbers; this being 193, and that shown by experience 160, we see that the formulas enunciated may be regarded as sufficiently demonstrated. We can also prove it in another way by calculating what resistance in u would equalise the two coils. This resistance is shown by the formula

$$u = \frac{R\,(t'\,H - t\,H')}{t\,(R + H') - t'\,(R + H)},$$

which gives, under similar conditions, 224 kilometres of telegraph wire, and the experiments give 200 kilometres.

3rd. The deduction which shows that the depth of the magnetising coils of an electro-magnet should be equal to the diameter of the cores which they cover, may easily be verified. In order to demonstrate this experimentally, I took three electro-magnets with bobbins of the same length but very different diameters.

One of these magnets had a diameter of 2 centimetres, another of 1 centimetre, and the third of ·65 centimetre. I applied one pole only to the balance, and each bobbin, covered with No. 16 wire, had 23 layers of 111 turns in each, or 2553 turns in all. Nothing of the thickness of paper even was introduced between the layers, and all the turns were pressed closely one against another, which gave these coils a uniform depth of 1 centimetre. The result was that the electro-magnet of which the core was 1 centimetre in diameter was the only one which answered to the maximum conditions previously given. This magnet had a resistance of 3200 metres, the large one of 5200 metres, and the small one of 2800 metres. The following (p. 54) are the results I obtained in passing through these different electro-magnets the current from a Leclanché battery containing from 1 to 3 cells, each having an internal resistance of about 400 metres, estimating the attractive force at a distance of 1 millimetre.

We see by this table that, for a like resistance of exterior circuit, and with a sufficient electric intensity, it is the electro-magnet of which the depth of the coil equals the diameter of the core, which has

Resistances of Exterior circuit, in metres.			Electro-Magnets.		
			Large, 2 centimetres.	Mean, 1 centimetre.	Small, ·65 centimetre.
			gr.	gr.	gr.
3 elements	1200 +	0 = 1,200	76	112	86
	1200 +	1,600 = 2,800	48	57	(44)
	1200 +	2,000 = 3,200	43	(50)	39
	1200 +	4,000 = 5,200	(28)	29	22
	1200 +	10,000 = 11,200	13	10	8
2 elements	800 +	0 = 800	38	57	46
	800 +	2,000 = 2,800	22	26	(20)
	800 +	2,400 = 3,200	20	(22)	16
	800 +	4,400 = 5,200	(14)	13	10
	800 +	10,000 = 10,800	7	5	4
1 element	400 +	0 = 400	12	18	15
	400 +	2,400 = 2,800	7	7	(6)
	400 +	2,800 = 3,200	6	(6)	5
	400 +	4,800 = 5,200	(5)	4	4
	400 +	10,000 = 10,400	3	2	2

the advantage; this advantage is found to be invariable when we compare the forces produced on exterior circuits of different resistance, adapted to the resistances of the magnets, the only case which has been discussed in the formulas. It is only when the electric force is so weak that the increase of the magnetic action with the diameter is not very palpable, that the maximum of the electro-magnet of 1 centimetre loses ground a little; and this is as it should be, for Muller's law, which supposes the attractive forces to be proportional to the diameter of the magnetic cores, is only true when these cores are magnetised to a pitch approaching that of their magnetic saturation; and by this word saturation must be here understood that magnetic state which the electro-magnet would preserve, if, instead of being iron, it were tempered steel magnetised. When the magnetic force developed is much below this point, it is on the electric intensity that the attractive force produced chiefly depends, and this is naturally greatest with the least resisting circuit. The figures in parentheses in the preceding table point out, in each of the three series of experiments made with different electric intensities, the forces corresponding to the maximum conditions with reference to the circuit, and these conditions were naturally established, supposing the resistance of the exterior circuit to be equal to that of the coil; for I started in the construction of my electro-magnets from a given depth of bobbin. But if I based my calculations on the maximum conditions relative to a given electro-magnet without considering its dimensions, the preceding law is still further verified, for the maximum resistances of the exterior circuit then become 1400 metres for the small magnet, 2600 metres for the large one, and 1600 metres for the other. Thus the attractive forces of these three electro-magnets are as follows :—

	With 3 Elements.	With 2 Elements.	With 1 Element.
	gr.	gr.	gr.
Electro-magnet of 1 centimetre	94	46	14
,, ,, ,, ·65 centimetre	70	41	13
,, ,, ,, 2 centimetres	50	25	8

CHAPTER VII.

EFFECTS OF A MORE OR LESS COMPLETE MAGNETIC SATURATION.

As I have said, the laws of electro-magnets are not as definite as might be desired, by reason of the various effects resulting from the state of saturation of their magnetic core; but this cause of perturbation is far from being as great as certain scientists would have it thought, and it interferes in a less proportion with the definite results deduced, than do the effects of polarisation with the calculated results of the laws of electric currents. At the time of my last investigations concerning electro-magnets, I wished to assure myself of the importance of this disturbing cause, and I undertook a great number of experiments which seem to me interesting to report here, for in consequence of the development that takes place every day in the application of electro-magnetism, it is, above all, important to be well grounded in the conditions for the good construction of magnets on which depends the success of these applications.

In the first part of this work I gave several general deductions which I had drawn from experience, but I did not insist on my experiments themselves, for they led to no law besides those given by Dub and Muller; but it is important that I should explain myself more explicitly in this respect; I will, therefore, begin by demonstrating that if we take 11 as the value of the coefficient *m*, by which the diameter of the magnetic core must be multiplied to find its length, so as to satisfy the different conditions of maximum relating to it, its attractive force always increases in proportion to its length. In fact, if we start with a given length of wire representing the resistance of the exterior circuit, and wind it on electro-magnets of different diameters so as to get a depth of coil equal to their diameters, their lengths must be different and calculated to satisfy the equations

$$\frac{2\pi b c^2}{g^2} = \frac{2\pi b' c'^2}{g^2} \quad \text{or} \quad b c^2 = b' c'^2$$

and then these lengths will be inversely proportional to the squares of their diameters. In this case the factor *m* is no longer constant and becomes proportional to the cube of the diameters; but then the law which supposes the E. M. F. proportional to the squares of the number of turns multiplied by the $\sqrt{\text{cube}}$ of the diameter, is no longer applicable, and we must then, to compare the forces, have recourse to the law which takes the latter to be proportional to the number of turns multiplied by the diameters of the cores, and the square roots of the lengths.

We have then for a like intensity of current—

$$\frac{A}{A'} = \frac{c^2 b^2 c \sqrt{b}}{c'^2 b'^2 c' \sqrt{b'}} = \frac{c^3}{c'^3} \frac{b^{\frac{5}{2}}}{b'^{\frac{5}{2}}} = \frac{c^3}{c'^3} \frac{c'^5}{c^5} = \frac{c'^2}{c^2} = \frac{b}{b'},$$

which shows that the forces are then in inverse proportion to the squares of the diameters, or proportional to the lengths, admitting, however, that the magnetic cores satisfy the desired conditions of saturation, so that the laws of Dub and Muller may apply.

To ascertain the modifications in the laws we have been considering, caused by the more or less complete saturation of a magnet in conjunction with its length and diameter, I wound on three different bar magnets of 8, 7, and 6 millimetres in diameter respectively, a similar length of wire (No. 12 of ·059 millimetre including the insulating covering). The length of this wire was 71·47 metres, equalling in resistance 722 metres of telegraph wire, or 7·22 ohms; and I calculated the length of my bobbins so that the depth of the coils was always equal to the diameter of the magnetic core. These bobbins had consequently lengths of 5·9, 7·7, and 9·8 centimetres respectively, and the numbers of turns were 1470, 1677, and 1842. The following are the results obtained with a battery of from 1 to 4 elements (latest type, Leclanché), the resistance of each cell being 113 metre, or 1·13 ohm. The attractive force was measured at a distance of 1 millimetre.

Resistances of Exterior Circuit, in metres.		Magnet of 9·8 centimetres.	Ratio of Magnetic Force.	Magnet of 7·7 centimetres.	Ratio of Magnetic Force.	Magnet of 5·9 centimetres.
		gr.		gr.		gr.
4 elements	452 + 0 = 452	225	1·018	221	·9210	240
	452 + 300 = 752	151	1·049	144	·9660	149
	452 + 400 = 852	134	1·0635	126	·9770	129
	452 + 1000 = 1452	73	1·0900	67	1·0307	65
	452 + 2000 = 2452	36	1·161	31	1·0333	30
	452 + 3000 = 3452	21	1·235	17	1·0630	16
	452 + 4000 = 4452	14	1·272	11	1·1000	10
3 elements	339 + 0 = 339	169	1·0432	162	·953	170
	339 + 300 = 639	109	1·0792	101	1·010	100
	339 + 400 = 739	95	1·092	87	1·0117	86
	339 + 1000 = 1339	50	1·163	43	1·0238	42
	339 + 2000 = 2339	23	1·2105	19	1·0555	18
	339 + 3000 = 3339	13	1·300	10	1·000	10
	339 + 4000 = 4339	8	1·333	6	1·000	6
2 elements	226 + 0 = 226	100	1·111	90	1·000	90
	226 + 300 = 526	60	1·154	52	1·0196	51
	226 + 400 = 626	52	1·1818	44	1·0232	43
	226 + 1000 = 1226	26	1·3000	20	1·000	20
	226 + 2000 = 2226	11	1·375	8	1·000	8
	226 + 3000 = 3226	6	1·500	4	1·000	4
	226 + 4000 = 4226	4	2·000	2	1·000	2
1 element	113 + 0 = 113	35	1·207	29	1·036	28
	113 + 300 = 413	19	1·2666	15	1·071	14
	113 + 400 = 513	16	1·3333	12	1·0909	11
	113 + 1000 = 1113	7	1·1666	6	1·200	5
	113 + 2000 = 2113	2	1·0000	2	2·000	1
	113 + 3000 = 3113	1	1·0000	1	2·000	0
	113 + 4000 = 4113	0	1·0000	0	2·000	0

These results show that in conformity with theory, the longest electro-magnet is usually the strongest, but in a ratio which is only in proportion to the lengths when the diameters are approximately equal, and for a certain intensity of current which probably corresponds with the point of magnetic saturation.

This intensity for the magnets of 9·8 and 7·7 centimetres, varied between ·001082 and ·001030, and between ·00143 and ·00113 in the four series of experiments; but in comparing the electro-magnets of 9·8 and 5·9 centimetres, we do not find this proportion for any intensity of current, and the ratio of force is always less than the length. Further even, it will be observed that the strength of these electro-magnets is so subject to the state of magnetic saturation of the cores, that for high intensities or low resistance circuits, it is the short, thick magnet which has the advantage. On the contrary, the preponderance of the long thin magnet becomes more and more marked as the electric intensity diminishes; whether this diminution arises from fewer cells being employed, or from higher resistances being introduced into the circuit, and we may judge of the importance of these variations from the proportions in the second and fourth columns in the preceding table. It will be understood that this should be the case, as for high electric intensities the diameter of the long electro-magnet is not proportional to these intensities, and its point of magnetic saturation is passed when it is hardly reached by the short, thick magnet. On the other hand the advantage of the small diameter is explained, when we consider that the magnetic mass of the bar being great enough to correspond with the electric force developed, it receives the full benefit of the greater number of turns in its magnetising coil. It may then be concluded that the dimensions to be given to an electro-magnet, must essentially depend upon the electric force which has to act upon it, and on the resistance of the circuit in which it is introduced. When the circuit is long and the electricity of low intensity, they should be long, and small in diameter; when, on the other hand, the circuit is short and the electricity of high intensity, the core should certainly be of large diameter. We also arrive at this deduction from the formula $c = \dfrac{E}{\sqrt{R}} \cdot 173$ which I have given for calculating the diameter to give to an electro-magnet according to the conditions in which it may be employed.

As to the influence of the magnetic saturation of the cores, it is somewhat difficult to formulate precisely. Joule, de Haldat, Muller and Robinson long ago discovered that, at the commencement of the action of the current, and when the magnetic state of the iron is far from the point of saturation, the attractive force, instead of increasing as the square of the intensity of the current, increases in a much more rapid proportion, which may be beyond the third or even the fourth power of this intensity; but they also showed that as the electro-motive force develops, this proportion diminishes rapidly, till the point of saturation is reached, when it remains some instants stationary, and then begins to decrease beyond the point of saturation until it becomes simply proportional to the intensity of the current.

If this be the case as the magnetic force is developed, it must be the same when this force, being capable of complete development, is produced by cores of which the dimensions for a given electric intensity receive a different magnetic saturation; and we may judge by the results of the experiments cited above of the truth of this deduction. In fact, if we take the ratios of force of each of the three electro-magnets in question when they are excited by different electric intensities, we find that for each of them there is an intensity with which the force increases as the square of the intensity, and above or below which it increases in a greater or less proportion. It will also be noticed that this limit of intensity varies according to the dimensions of the electro-magnet. Thus, on compiling the following table from these proportions, we see, firstly, that the electro-magnet B best satisfies the law of the squares of electric intensities; secondly, that the large magnet C gives a more rapid proportion; thirdly, that the magnet A of small diameter gives a slower proportion; fourthly, that the ratio of forces for these three electro-magnets is so much more rapid compared with what it should be, according to the law of the squares of the intensities, as these intensities become weakened, whether this weakening results from lessening the number of cells in the battery or from an increase in the resistance of the circuit. It is true that these ratios, notwithstanding their great value, more nearly approach the squares of the intensities than the cubes, but it will be noticed that the three electro-magnets have somewhat similar diameters, and are very nearly in the same condition of saturation, and that the forces have been measured at their maximum magnetisation.

It is evident that it would not have been the same with electro-magnets of diameters differing more widely, or if the forces could have been measured instantaneously at the moment of magnetisation, and during a very short closing of the circuit. In this case we might have found a ratio even higher than 85·600, the cube. It is besides on this principle

Ratio of the Squares of the Intensities.	Ratio of the Magnetic Force of		
	Magnet A of 9·8 centimetres.	Magnet B of 7·7 centimetres.	Magnet C of 5·9 centimetres.
4 elements 1·576	1·49	1·54	1·61
1·797	1·68	1·75	1·86
3·429	3·082	3·30	3·70
7·309	6·25	7·13	8·
12·640	10·71	13·	15·
19·423	16·07	20·09	24·
3 elements 1·645	1·55	1·60	1·70
1·896	1·78	1·86	1·97
3·773	3·38	3·77	4·05
8·323	7·35	8·52	9·44
14·650	13·	16·20	17·
22·753	21·12	27·	28·33
2 elements 1·733	1·66	1·73	1·76
2·022	1·92	2·04	2·09
4·222	3·84	4·50	4·50
9·670	9·09	11·25	11·25
17·343	16·66	22·5	22·5
27·242	25·	45·	45·
1 element 1·847	1·84	1·93	2·
2·187	2·19	2·42	2·55
4·829	5·	4·83	5·60
11·527	17·50	14·50	28·
21·094	35·	29·	
33·529			

that the action of slow-working magnets in some printing telegraphs is based, and which are only slow-working because their mass being relatively great, they take a certain time to magnetise. However, the foregoing is the table to which I have referred.

We may deduce from the preceding results, that the law of the proportion of attractive force to the square of the intensity of the current is only true within certain limits and under certain conditions and that electro-magnets, on a circuit which is rapidly made and broken, are more or less exempt therefrom. And as this is the case in which it is most important to ascertain what are the best conditions, it has appeared to me that the matter should be restudied from this point of view, not to fix new arbitrary conditions of maximum, but to ascertain in what way those already deduced must be modified to suit any particular case.

The question to resolve is this: When the electro-magnetic force increases in a greater ratio than that of the square of the intensity of the current, as, for instance, as the cube of this intensity, should the resistance of the magnetising coils be greater or less than that of the exterior circuit?

To solve this question we have only to change in the formula giving the value of the electro-motive force, the indices of the quantities in proportion with this quantity; in a word, to transform I^2 into I^3. We thus obtain a reduced expression

$$8f^2 \pi b a (a + c) = 4 q R g^4 ;$$

when we vary g, which gives for conditions of maximum

$$\frac{\pi\,b\,a\,(a+c)}{g^2} = \frac{q\,\mathrm{R}\,g^2}{2f^2}, \text{ or, } \mathrm{H} = \frac{\mathrm{R}}{2};$$

and when we vary a these conditions will answer to the equation

$$\frac{q\,\mathrm{R}\,g^2}{f^2} = \frac{\pi\,b\,a\left(2\,a + \dfrac{c}{2}\right)}{g^2}.$$

These two equations show that with the force proportional to the cube of the intensity, the coils must always have half the resistance of the exterior circuit, when we take g as variable, and in the proportion of $2\,a + \dfrac{c}{2}$ to $a + c$ when a is variable. We may thus conclude that on circuits where the interruptions of current are rapid, the resistance of the electro-magnet coils must be as much less as the closing of the circuit is of short duration; and it is for this reason, as well as on account of the defective insulation of telegraph lines and the development of extra currents, that Mr. Hughes first, and the electricians of the telegraph department afterwards, have considerably reduced the resistance of electro-magnets introduced on long circuits.

It is also of interest to know how the resistance of an electro-magnet should be modified, when, the point of saturation being passed, the forces increase less rapidly than the squares of the intensities. If

F

we suppose that this increase is in a ratio simply proportional to the intensity of the current it will be seen that there will be no maximum possible, and therefore we may advantageously increase the resistance beyond the limits which have been assigned.

If we now endeavour to ascertain how the conditions of maximum, as regards the depth of the magnetising coil, may be modified in consequence of the complete saturation of the core, we shall see that we may then increase this depth, which might with advantage be double the diameter of this core, if the force increased as the cube of the intensity of the current. The maximum conditions of the formula

$$\frac{E^3 \, t^2 \, c}{\left[\dfrac{\lambda \, \pi \, b \, a \, (a + c)}{g^2} \right]^3}$$

give, in fact, for conditions of maximum, taking c as variable, $a = 2\,c$. No maximum is, however, possible, if we consider that the forces are simply proportional to the intensities of the current.

CHAPTER VIII.

CONDITIONS FOR THE GOOD CONSTRUCTION OF ELECTRO-MAGNETS.

IT is not only necessary to calculate accurately the dimensions to give to an electro-magnet, but also to employ it in such a way that it may most efficiently develop its proper force, taking into account any secondary action, upon which its useful application may greatly depend. It is this which we will now consider on the basis of the different results I have obtained from very numerous experiments made by me during the last twenty years and more.

I. *Conditions in reference to exterior actions.*— Descartes, and several other scientists after him, showed that if a mass of iron be applied to one pole of a bar magnet, the attractive force of the other is increased in proportion to the size of the iron. Dub, in applying this phenomenon to the laws of electro-magnets, stated that this increase of force arises from the fact that the magnetic core is thereby lengthened, and that the forces ought to increase in consequence: but this increase, according to him, would only be as the square roots of the lengths, and according to my experiments this increase is infinitely greater, since I have often found the forces tripled. I have also found that this increase may be produced with masses of iron of all shapes, which do not materially increase the length of the core, and even when the iron is not in contact with the magnetic core. I found also that the development of the magnetised surface has great influence on this phenomenon, and I have in consequence come to the conclusion that this increase of energy results principally from magnetic condensation towards the iron, having the effect of stimulating the opposite free pole of the magnet, since the opposite poles of a magnet must correspond in force.

I also directly demonstrated this deduction by placing on a long rod of soft iron, magnetising coils of different lengths but composed of equal lengths of wire, and I found that it was the coil that entirely covered this rod that produced the least attractive force although it had the greatest number of turns. The coil which gave the best result was the one which only covered one-third of the rod. When, however, these same coils covered iron cores of the same lengths as themselves, the result was quite different; the force then increased with the length of the coils in a proportion less rapid than that of the lengths, but which, from my experiments, seemed to me to be in that of an arithmetical progression, when the lengths increased as a geometrical progression; and the terms of this progression seemed to increase in proportion to the number of cells employed. I only give these results, however, as applying to a particular case.

The effects which I have described above, explain the cause of the relatively great force developed by horse-shoe magnets, with a bobbin on only one of the arms; electro-magnets which I was the first to employ, and to which I gave the name of one-legged (*boiteux*) electro-magnets.

II. *Conditions in reference to the form and application of the armature.*—The results of my experiments in this respect may be numbered as follows:—

1st. The attractive force of any electro-magnet is greater in proportion as the surface of its armature, which directly receives the magnetic influence, is extended, as it is thus brought into better relation with the magnetic energy of the electro-magnet.

2nd. It follows that the attractive force of a horse-shoe electro-magnet at a distance, is greater with a prism-shaped armature placed lengthwise before the poles, than with the end presented, although the reverse is the case when the attractive force is exerted on contact. For an attraction at a distance the effects produced may be in the proportion of 59 to 92.

3rd. Armatures moving at an angle with the line of the poles of an electro-magnet, i.e. working on a joint near one of the poles, are much more effective than armatures moving parallel to this line or fitted cross-wise to a rocking lever. This advantage is especially noticeable with the one-legged electro-magnets, and the force varies in the proportion of 125 to 64.

4th. Prismatic armatures are attracted in proportion as the surface is extended, but the form of this surface has an immense influence on the attraction, on account of the mean distance of all the points which are under the influence of the magnet, which distance may vary very much with this form. Thus a cylindrical armature of the same surface as a square one is attracted with much less force that the latter, in the proportion of perhaps 85 to 44.

5th. In like manner the lateral attraction of an electro-magnet whose ends protrude a little beyond the bobbins, is infinitely less than the normal attraction, i.e. that in line with the axis of the poles, in the proportion of 33 to 18.

6th. Armatures formed of permanent magnets do not increase the attraction except at a distance, and when moving parallel to the line of the poles. In other cases, the reverse obtains, as the magnetic action on the steel is much less than on iron.

7th. The attractive force obtained by a momentary closing of the circuit for a similar distance, is greater than that obtained by the same current continuously applied in the hope of overcoming the resistance. This is on account of the *vis viva*, and of the effects of polarisation of the battery. The proportion of these forces is as 136 is to 95.

8th. When the attractive force of an electro-magnet is divided over several armatures, the total attractive force is increased, but the individual force of each is diminished in proportion to their number. This increase, however, is only shown up to a certain limit, which is attained when the mass of the armature equals that of the electro-magnet.

9th. The attractive force of an electro-magnet and armature which have never been used, is greater for a given electric force, than that of the same magnet and armature after being strongly magnetised, and to obtain from the same magnet and armature a nearly equal force to that first obtained, the current must be reversed; still this greater strength is then only to be found on first closing the circuit.

10th. The attraction at a distance is less, when from some cause the first closing of the circuit has not been followed by a complete attraction of the armature; this is explained, as also the preceding result, by the effect of the residual magnetism.

11th. The repulsive force developed by electro-magnets on a magnetised armature is very far from corresponding with the attractive force which may be obtained by reversing the poles of the magnet. This fact, recognised in the earliest researches on magnetism, and fully investigated by Mussembroeck and the Abbot Nollet, is explained by the fact of the magnet acting inductively, thus tending to develop in the armature a polarity opposite to itself. In attraction, this influence increases the result, whereas in repulsion it has the contrary effect.

12th. When the iron cores protrude beyond the magnetising coils their strength is diminished, but if iron plates are fitted round the cores, the attractive force is increased, and the maximum effect is obtained when the distance between the two rings is about a quarter of the distance between the poles. This is occasioned by the greater attractive surface presented, which then nearly corresponds with that of the armature.

13th. According to the experiments of Dub, the best results are obtained when the different parts of the electro-magnet (arms of the magnet, breech and armature) are equal in mass.

14th. This conclusion is of course subordinate to the conditions of application, for it is certain that if we wish to obtain a quick movement of the armature we must have a light one; but we may make up for it by providing the poles with iron rings.

III. *Conditions in reference to the magnetic mass.*— The conditions of force of electro-magnets with reference to their diameter and their degree of magnetic saturation, may, in certain circumstances, be contradictory one to the other; for, if we gain in force by increasing the diameter of the magnetic cores, we lose when, their diameter being too large, the cores are not saturated, and this is in consequence of secondary reactions exercised on the outside of the cores by the central part, which is inert and plays the part of a second armature. It has been tried to obviate this inconvenience by employing tubular cores, which at the same time facilitated rapid magnetisation and demagnetisation; but with this system a considerable lessening of the attractive force was the result, and it was abandoned for the moment. In consequence, I undertook the study of the conditions of force of this sort of electro-magnet, and after numerous experiments recorded in a memorandum presented to the Académie des Sciences in 1862, I arrived at the following conclusions :

1st. The greater force of electro-magnets with solid cores is not in consequence of their greater mass of metal, but chiefly depends on the disposition of the polar surfaces with regard to the armature.

2nd. If the polar extremity of a tubular core is provided, inside the tube, with an iron stopper, which may be very thin, the force of the electro-magnet is nearly the same as with a solid core ; but this is not the case if we increase the polar surface by surrounding the end of the tube with an iron ring, which shows that it is not so much the increase of the polar surface as its disposition which acts upon the attractive force.

3rd. If we consider that the ring acts as an armature to the detriment of the external attraction, as when an electro-magnet acts at the same time on several armatures, while with the iron stopper we have a concentration of the magnetic effects produced by the different parts of the internal walls of the tube, as much below the stopper as laterally, we can understand that there must be a considerable difference of force in the two cases ; and we may easily see the effect of the magnetic concentration produced in the second instance, by the projection of the iron stopper out of the tube at the moment of magnetisation, provided that the stopper is not fitted in too firmly.

4th. It results from these effects, that, to usefully employ tubular electro-magnets, they should be provided at the polar ends with iron stoppers of a thickness equal at least to that of the tube. By this arrangement an increase of force is gained, which in telegraphic electro-magnets might be in the proportion of 25 to 38.

As to the thickness of the tubular cores, it should be in proportion to the magnetising current. Hughes found by experience that it should be a quarter of the diameter of the tube; but I have found, by experiment, that for electro-magnets of large diameter, this thickness may be reduced in still greater proportion. In a memorandum inserted in my 'Recherches sur les meilleurs conditions de construction des électro-aimants' (page 112), I treated this question at length, and showed that the diameter c' of the tube should be expressed by

$$ c' = c \sqrt[3]{\frac{x^3}{4\,(x-1)}}, $$

c being the diameter of a solid iron core, susceptible of being magnetised to saturation, and x the divisor of c' to represent the thickness. This value of x may in some cases be as great as 7.

5th. The disadvantages of solid electro-magnets result chiefly from the effects of residual magnetism, and from magnetic condensation, but they may be considerably reduced by magnetically isolating the arms of the electro-magnet from the breech by means of rings of copper, or by cutting in two the breech itself by a thickness of copper.

CHAPTER IX.

ON THE BEST GROUPING OF THE CELLS OF A BATTERY.

As it is important to arrange an electro-magnet in such a manner as to place it in its maximum conditions with regard to a given circuit of resistance, so we should endeavour to combine the cells of a given battery so as to make it of most use on a given circuit with a given electro-magnet. Ohm has established for this particular case a formula, generally given in electrical treatises, but not very applicable in practice, and this has induced me, since 1860, to endeavour, by another method, to ascertain the best way of grouping the elements of a battery. What is most important indeed to know, is what is the most advantageous mode of grouping the elements of a given battery to correspond with a known exterior circuit of resistance R, and what number of groups there should be, also the number of elements in each. Thus, for a battery with a number n of elements, the point in question is, whether for a circuit of resistance R, it would be better to form the elements of this battery into four groups in tension, each consisting of three elements in quantity, or six groups of double elements, or two groups of six elements, &c. Calculation will give us the desired answer.

Let us suppose that we have a battery composed of n elements; we will take a for the number of groups which should be joined for tension—i.e. by opposite poles—and b for the number of elements joined for quantity in each group—by similar poles. If we want an expression for the intensity of the current of each of these groups, by representing by r the resistance of one element, by R the resistance of the circuit, and E the electro-motive force of a single element, we have, according to Ohm's formula with respect to the grouping for quantity

$$I = \frac{E}{\dfrac{r}{b} + R} \, ,$$

and if we join for tension all these groups, of which the number is represented by a, we have

$$I' = \frac{a\,E}{\dfrac{a\,r}{b} + R} \text{ or } I' = \frac{a\,b\,E}{a\,r + b\,R}.$$

This formula, as well as that of Ohm, is susceptible of maximum, for $a\,b = n$ and $b = \dfrac{n}{a}$; so that the equation becomes

$$I' = \frac{n\,E}{a\,r + \dfrac{n}{a}\,R} = \frac{E}{\dfrac{a\,r}{n} + \dfrac{R}{a}} \, ,$$

and the denominator being, with respect to a, considered as variable, $\dfrac{r}{n} - \dfrac{R}{a^2}$; we obtain the maximum

conditions by making $\dfrac{r}{n} - \dfrac{R}{a^2} = 0$, or $a = \sqrt{\dfrac{n\,R}{r}}$,

and as $a = \dfrac{n}{b}$, we can also obtain the value of b,

which is $\sqrt{\dfrac{n\,r}{R}}$ or simply $\dfrac{n}{a}$.

We could, however, perfectly demonstrate this deduction by simple reasoning, without the intervention of differential calculus, as I have shown in the first volume of my 'Exposé des applications de l'électricité' (page 158).

There result from these conditions of maximum some rather important consequences:—

1st. That when a battery is placed in its maximum conditions with respect to a given exterior circuit of resistance R, this resistance is equal to that of

the battery; and from the equation $a = \sqrt{\dfrac{n\,R}{r}}$ we

get $R = \dfrac{a}{b}r$, and the second side of this equation represents the resistance of the battery.

2nd. That we can know the most advantageous method of arranging the elements of a battery, starting from a given intensity I, provided this intensity does not answer to a resistance R greater than $n\,r$. And as, in its maximum conditions, we have

$$\frac{n\,E}{a\,r + b\,R} = \frac{n\,E}{2\,a\,r} = \frac{n\,E}{2\,b\,R},$$

finally we have from this

$$a = \frac{2\,I\,R}{E} \text{ and } b = \frac{2\,I\,r}{E},$$

very convenient formulas and extremely simple.

In the different works I have undertaken on this subject, which were successively presented to the Académie des Sciences in June and August 1860, and in September 1869, I have pointed out the limits of the resistance of the exterior circuit, between which it is most advantageous to employ such and such a mode of grouping; and these limits, for the coupling of b elements joined for quantity are $\dfrac{n\,r}{(b-1)\,b}$ and $\dfrac{n\,r}{(b+1)\,b}$, that is to say, for double elements; these resistance limits of R are $\dfrac{n\,r}{2}$ and $\dfrac{n\,r}{6}$ for triple elements; $\dfrac{n\,r}{6}$ and $\dfrac{n\,r}{12}$ for quadruple elements; $\dfrac{n\,r}{12}$ and $\dfrac{n\,r}{20}$, &c.; and I have shown also, as Ohm had established, that the resistances R corresponding to the maximum of the different sorts of coupling are represented by $\dfrac{n\,r}{b^2}$, that is to say by $\dfrac{n\,r}{4}$ when the battery is arranged in double elements, by $\dfrac{n\,r}{9}$ when it is arranged in triple elements, by $\dfrac{n\,r}{16}$ when in quadruple elements, &c., &c. I thus show that the limits of resistance of the exterior circuit, to give the same intensity with double, triple, quadruple cells, &c., compared with cells coupled up for tension or for quantity, correspond on the one hand with the half, third, or fourth of the total resistance of the battery, and on the other, with the half, third, or fourth of the resistance of a single cell.*

Thus we shall know by means of the preceding calculations, that, if we have a battery of 20 Daniell cells, the resistance of the exterior circuit which would necessitate its arrangement in double elements, will be between 10 kilometres and 3333 metres

---

* These different deductions are obtained by successively equalising the different expressions representing the intensity of the current furnished by the different groupings of the battery, and extracting the value of R. We may even employ this means to demonstrate without the assistance of the differential calculus, the conditions of maximum of the formula, for, supposing b and b' to represent the number of cells in quantity in two nearly similar groupings, we find on equalling the two expressions,

$$a\,r + b\,\mathrm{R} = a'\,r' + b'\,\mathrm{R}, \quad \text{or,} \quad \mathrm{R}(b - b') = r\left(\frac{n}{b'} - \frac{n'}{b}\right),$$

whence

$$\mathrm{R} = \frac{n\,r}{b\,b'},$$

which may be transformed into

$$\mathrm{R} = \frac{n\,r}{(b-1)\,b}, \quad \text{or,} \quad \mathrm{R} = \frac{n\,r}{(b+1)\,b},$$

according to whether we compare the two groupings in one way or the other. But if we so approximate one to the other that $b' = b$ supposing that the battery corresponds with the maximum arrangement, the preceding expression becomes $\mathrm{R} = \frac{n\,r}{b^2}$, and thence we arrive at the equations expressing the maximum values of a and b given above.

of telegraph wire. When the resistance is between 3333 metres and 1666 metres, we shall gain by arranging the cells three in series, and between 1666 and 1000 metres we ought to group them in fours.

From what we have said above, it will be seen that nothing is easier than to calculate directly what arrangement it will be best to make of a given number of cells, to correspond with a given circuit of resistance; since if we divide successively by 4, 9, 16, &c., the total resistance of n elements, we find which is the nearest to the given external resistance. The figure of the divisor gives the square of the number of elements in each group, and the number of groups is obtained by dividing the total number of elements by the number in quantity in each group.

These calculations, as may be seen, are exceedingly easy and within the reach of any intelligence, and I have always been astonished that inventors have not given them more attention, and thereby avoided loss of time and money in useless researches, when they might at once have ascertained the maximum conditions obtainable.

Some people have objected that these calculations cannot be precisely correct, unless we can get whole numbers for a final result, as the elements of a battery cannot be divided into fractions; but in electricity we must be content with approximate data. Certainly the elements of a battery cannot be divided up, and owing to that we are rarely able exactly to satisfy the conditions of theoretical maximum ; but we have a guide, and it the nearest whole number which must be made use of. Thus, for example, in the case mentioned in Chapter V. we find that the value of a is $3 \cdot 074$; and we must evidently take 3 groups in tension as the nearest we can get; and as 8 divided by 3 gives $2 \cdot 7$, we must put 3 elements in quantity as the best arrangement we can make.*

The formulas which we have given above enable us also to compare the forces of the different electric generators. There are, however, in these calculations, certain considerations which not having been looked at from the same point of view by different scientists, have led to regrettable disagreements. Thus, according to Jacobi, two different batteries cannot be compared unless they are both placed in their maximum conditions; but for this it would be necessary for the external resistance of their circuits to be equal to their internal resistance, and therefore their intensity would be represented by $\dfrac{n \, \mathrm{E}}{2 \, r \, a}$ or $\dfrac{n \, \mathrm{E}}{2 \, b \, \mathrm{R}}$.

And if, for the battery we are taking as the basis of comparison, we take $n = 1$, reducing the preceding

* When the results given by calculation do not correspond with whole numbers, the conditions of maximum cannot evidently be exactly fulfilled by taking the whole numbers nearest to them. Gangain mathematically examined this phase of the question, but only arrived at undecided and complicated solutions, for he found that sometimes a problem had two and even more solutions, while others had only one. (See *Annales Télégraphiques* for January 1861.) At any rate the method described above is sufficient to obtain a practical maximum arrangement of the cells of a battery in any given case.

formula to $\dfrac{E}{2\,r}$, we have, supposing the intensities equal,

$$\frac{n\,E}{2\,a\,r} = \frac{E'}{2\,r'} \quad \text{and} \quad \frac{n\,E}{2\,b\,r} = \frac{E'}{2\,r'},$$

which may also be written,

$$\frac{b\,E}{2\,r} = \frac{E'}{2\,r'}, \quad \text{and} \quad \frac{a\,E}{2\,R} = \frac{E'}{2\,r'},$$

and from which we may obtain the values of a and b, if we suppose the resistance R equal to the resistance of the test cell. We have then,

$$a = \frac{E'}{E} \quad \text{and} \quad b = \frac{E'}{E} \times \frac{r}{r'},$$

and the total number of elements is then given by multiplying a by b.

In this manner we find that 5 Bunsen cells equal 9 Daniell cells of 11 times the surface, or in other words that 20 Daniells of ordinary size are equal to one Bunsen. We also find that 3 Bunsen cells equal in intensity 4 sulphate of mercury cells with $3\frac{1}{3}$ the surface of the Bunsen, or that to equal one Bunsen we require 5 sulphate of mercury elements.

If the idea of Jacobi in giving the preceding system of comparison has been well understood, it will be seen that he endeavoured to avoid giving to the exterior resistance R a definite value; but in fact this is represented by that of the battery serving as basis for the comparison, and in the preceding examples it is equal to 153 metres of telegraph wire. And it will be seen that as this resistance is increased or diminished, the figures given above may be singularly modified; for the value of R may then become less and less important, or may become more and more so, in the value of the denominator of the formula giving the intensity of the current. This, then, has caused many scientists to criticise this system. We might take, as a basis for the resistance R, the mean of the resistance of the batteries to be compared, but this would not advance matters much, and it could only be applied to short circuits. In this case we might start with the intensity I of the test battery and obtain the values of a and b of the battery composed according to the formulas $a = \dfrac{2\,I\,R}{E}$ and $b = \dfrac{2\,I\,r}{E}$, which in the case of batteries of Daniells and Bunsens, for a resistance R of 153 metres, would give

$$I = 36 \cdot 55,$$

$$a = \frac{2 \times 36 \cdot 55 \times 153}{5973} = 1 \cdot 87,$$

and

$$b = \frac{2 \times 36 \cdot 55 \times 931}{5973} = 11 \cdot 4,$$

which bring us back to the conclusions previously arrived at, for then R is equal to the resistance of the test cell; but if R equals 1000, I becomes equal to $9 \cdot 64$; $a = 3 \cdot 2$ and $b = 3$, and we see that instead of employing 20 Daniell cells, we have only to employ 9 to be equal to the Bunsen cell.

TABLE I.

VOLTAIC CONSTANTS.

Batteries in most general use.	Electro-motive force. E.	Internal resistance in metres.		Specific value of E.
		With high external resistance.	With low external resistance.	
Daniell cell (sulphate of copper solution) French telegraph pattern. Small size	5973	931	180	1·00
Bunsen, with amalgamated zinc, shortly after charging. Same size as foregoing ..	11123	153	57	1·86
Delaurier cell, with chromic acid and solution of common salt. Same size as foregoing	12413	366	,,	2·08
Bichromate of potash, one liquid (Chutaux). Large size	11400	600	,,	1·91
Duchemin cell, perchloride of iron and solution of salt. Large size	9640	942	,,	1·61
Sulphuric acid and water cell, the two liquids separated. Small size	8547	880	,,	1·43
Marié-Davy cell, sulphate of mercury and water, the two liquids separated. Small size..	8192	550	,,	1·37
Leclanché cell, peroxide of manganese and solution of sal-ammoniac. Large size	7529	400	,,	1·26
De la Rue's fused chloride of silver and zinc. Diminutive size	5596	748	,,	·94
Sulphate of lead cell (Prudhomme). Same size as ordinary Bunsen	3301	880	225	·55

TABLE II.

Calculated Values for the Wires of Electro-Magnets.*

	Wires (French gauge).	Approximate B. W. G.	$\frac{g}{f}$ or s	g	f	f^2	$\frac{s^2}{s'^2}$	g^2
	No.	No.	metres	metres				
Silk Covered.	32	34	·00014	·00023	1·6428	2·6988	816·3	·0000000529
	28	32	·00022	·00033	1·5000	2·2500	330·6	·0000001089
	24	31 full	·00027	·00040	1·4814	2·1934	219·5	·0000001600
	20	28	·00035	·00048	1·3714	1·8796	130·6	·0000002304
	16	27	·00040	·00055	1·3750	1·8906	100·0	·0000003025
	12	25 bare	·00049	·00065	1·3265	1·7582	66·6	·0000004225
	P	24 full	·00058	·00077	1·3275	1·7609	47·5	0000005929
Cotton Covered.	1	23 bare	·00060	·00122	2·0333	4·1331	44·4	000001488
	2	22 ,,	·00070	·00134	1·9143	3·6634	32·6	·000001795
	3	21 ,,	·00080	·00146	1·8250	3·3306	25·0	·000002132
	4	20 full	·00090	·00158	1·7555	3·0800	19·8	·000002396
	5	19 bare	·00100	·00172	1·7200	2·9584	16·0	000002958
	6	19 full	·00110	·00184	1·6727	2·7989	13·2	·000003385
	7	18 bare	·00120	·00196	1·6333	2·6687	11·1	·000003842
	8	18 full	·00130	·00208	1·6000	2·5600	9·4	·000004326
	9	17 bare	·00140	·00220	1·5714	2·4670	8·2	·000004840
	10	17 full	·00150	·00232	1·5466	2·3901	7·1	·000005382
	11	16 full	·00170	·00254	1·4941	2·2320	5·5	·000006451
	12	15 bare	·00180	·00266	1·4777	2·1844	4·9	·000007075
	13	14 full	·00200	·00288	1·4444	2·0851	4·0	·000008294
	14	14	00210	·00300	1·4285	2·0392	3·6	·000009000
	15	13 bare	·00235	·00327	1·3915	1·9349	2·9	·000010690
	16	13 full	·00260	·00354	1·3615	1·8523	2·4	·000012532

* These diameters of the wire, with or without its covering, are not exactly the results that would be obtained by actually measuring the wire ; but these dimensions have been calculated from electro-magnets already made, whereby empty spaces in the coils and irregularities in the winding, which cannot be avoided, are brought into the values, and although the figures are in consequence not mathematically exact, the result is that we thus have much more correct data for practical purposes.

TABLE III.

TABLE OF THE RESISTANCES OF COPPER WIRE USED IN ELECTRIC APPLICATIONS.

Diameters in millimetres.	Lengths per kilogramme in metres.	Resistances per kilogramme in ohms.	Resistances per kilometre in ohms.
·02	355584	1803084	52317
·10	14369	30377	2113
·20	3614	1922	531
·30	1607	378	235
·40	902	119	133
·50	576	49	85
·60	401	24	59
·70	294	13	43
·80	226	7·5	33
·90	178	4·6	26
1·00	144	3·0	21
1·50	64	·59	9·2
2·00	36	·19	5·3
2·50	23	·078	3·39
3·00	16	·037	2·36
3·50	12	·020	1·72
4·00	9	·011	1·32
4·50	7	·0074	1·04
5·00	5·76	·0049	·84
5·50	4·71	·0033	·70

www.ingramcontent.com/pod-product-compliance
Lightning Source LLC
Chambersburg PA
CBHW051416200326
41520CB00023B/7260